SIX
WORDS

· · · ·· · ·· · ·· · ·· · · ··

SEMINAR HELD
IN A
PARALLEL UNIVERSE

SURENDRA KUMAR SAGAR

THE SEQUENCE

PREFACE

In the Indian *Upanishads* there is an equation "A = B"

or rather A •••• = B •• • •••

In the words of Schrödinger, "The personal self equals the omnipresent eternal self."

This equation was arrived at by intuition alone with the available knowledge/information at the time, which did not include knowledge about quantum physics, Einstein's relativity, or even Newton's laws.

We are now in the twenty-first century and our minds are halfway through the journey toward becoming "super minds." While it is true we have gained immense knowledge since the days of the Upanishads, we have also realized that for every question that gets answered several new questions pop up. It seems that the quantum of knowledge still to be gained keeps rising all the time.

Now, the above equation is unprovable. We all know that, but it answers many of the questions the new science is asking us. The closer we get to the "super mind" stage, the more we realize its potential of "making sense" and with it the realization that the universe itself perhaps makes sense.

This book shows how this equation leads to the "six words," the six words that show why or how this universe makes sense. These are not provable, too. Again this requires the initial condition (belief) to be that the universe makes sense.

But what do we mean when we say that the universe makes sense?

It means I am alive and conscious of the fact that the universe exists.

For, if I am not conscious, then where is the universe?

On second thought, where is the universe anyway?

What is the ratio of matter to emptiness?

I am thinking…Is this a universe of pure thought?

INTRODUCTION

The search for the ultimate truth can for all practical purposes be considered as the search for the "Probable Ultimate Truth."

Getting at the ultimate truth is not easy and it is expected that it will take a long, long time...perhaps millions of years.

But can our human race survive that long?

It looks as if an urgency has been created by "The Situation" on planet Earth, namely the very high probability of the human race extinguishing itself through self-destruction within the next three or four centuries, which is considered the danger zone here. The second law of thermodynamics is being played out to perfection by human civilization on planet Earth. Its entropy (disorder and chaos) is forever increasing.

It is a possibility that the knowledge of the Probable Ultimate Truth and its acceptance may provide us with a good chance to reverse the direction of this entropy, and enable us to survive the danger zone and come out unscathed.

Six Words is an autobiography that begins at the Big Bang, as I entered the universe in the form of a quark, and covers everything of consequence that happened to me from that time till now including my experience in a supernova. It deals with philosophy and attempts to arrive at a philosophical model that has a good chance to lead us toward that Probable Ultimate Truth.

These Six Words, as words, are simple. Put together these combine into a simple sentence. How do they lead to (or become) the Probable Ultimate Truth? This book is an exploration into that realization. If these words are announced as a matter of fact, they are unlikely to create an impact, let alone cause euphoria. "What wishful thinking..." "Too good to be true..." "Can't be true..." "Can never be proved..." These could be the expected reactions. An entire book needed to be written to convince the reader that these six words make sense, and that there is a good chance they might be true.

About one-fourth of this autobiography is made up of extracts from famous books, quotations, and statements by famous scientists and philosophers, who have immensely inspired me, and who I deeply admire. These scientists and philosophers and my deep appreciation of quantum physics and cosmology have led me to write this autobiography. However, I am neither a qualified quantum physicist nor a cosmologist, or a philosopher, and it was a daunting task to write on these subjects, merely by having read many books.

Quantum physics is absolutely counterintuitive. It defies common sense. We ask the question: "How can it be like that?" but there is no convincing answer. It just happens that way, and there is no explanation for it. Quantum particles move about here and there, individually, on their own, with complete disregard to the phenomenon of cause and effect. There is, however, a certain guiding force! I understood this guiding force, and with this understanding, I knew how to proceed.

Don't worry, though, I do not address the mathematics of quantum physics and very little concerning the mathematics formulation of the quantum theory. That part of quantum physics is meant for the specialists who intend to make use of the subject for practical purposes.

Our interest here is in the philosophy of quantum physics (and of course "cosmology"). We are more interested in understanding "What happens at the Quantum level?" It doesn't matter—for the time being—if it makes sense or not. We'll soon try to understand *how* it does happen.

So, I spread out all the books—I have plenty of them—on my study table and then prepared a sequential history of quantum physics as it developed over the years. I focused on the "giant leaps of imagination" experienced by the great scientists during that period when this branch of science took center stage in world history. Indeed, centuries later, when the important events of the first half of the twentieth century are written about, the development of quantum physics will be considered of greater significance than World Wars I and II.

Six Words deals with complex science and philosophy. However, don't worry, just read the complex parts once or twice and accustom yourself to them. Do remember, they are complex for me, too.

Chapter One

STANDARD MODEL OF PHILOSOPHY

"There has been one and only one sequence of events from the Big Bang through the entire history of the universe with all its galaxies, star formations, supernovas, and planetary systems including our own solar system, Earth, and the evolution of life on Earth, the human race, your arrival in the world and immediate containment in a speck of jelly, your growing up, continuous interactions of the world on you, which have made you precisely what you are up to the present moment, ultimately leading you to read the following sentence."

"There has been one and only one sequence of events from the Big Bang through the entire history of the universe with all its galaxies, star formations, supernovas, and planetary systems including our own solar system, Earth, and the evolution of life on Earth, the human race, your arrival in the world and immediate containment in a speck of jelly, your growing up, continuous interactions of the world on you, which have made you precisely what you are up to the present moment ultimately leading you to read the following sentence."

"There has been..."

And so on...A regression...

There is no way you could have been any different from what you are and there is absolutely no way you could have avoided reading the above.

If the entire duration of time from the Big Bang to the present (approximately fourteen billion years) is considered as twenty-four hours with the Big Bang at time zero, human beings arrived only at fifty-four seconds after 11:59 p.m., just six seconds back. Already we are talking of global warming, depletion of resources, population explosion, and possible nuclear wars. Several analysts predict the end of human civilization in less than three hundred years...unless, good sense prevails. At the scale indicated this works out to 0.0018 seconds.

Our enormous universe may last for hundreds of trillions of years. But, has the universe been created to enable intelligent life to exist for such a short time, on one planet of just one stellar system out of several trillions of stellar systems? If that sounds unlikely, life must exist on several planets in several star systems all over the universe. Though ours may be among the top one thousand technically advanced civilizations of the Milky Way, we are not likely to reach the top ten. We, like most others, will self-destruct to extinction...unless good sense prevails.

What does good sense entail? What is leading us to our imminent self-destruction? Are the causes our different, narrow points of view, our conflicts? Is the cause of self-destruction our philosophy of life that leads us to follow different religions, which come into conflict with each other? Is the cause science conflicting with these different religions? Is a universal faith—a convergence of religions into one and that faith's convergence with science the answer? This book is an autobiography that begins at the Big Bang and avoids a sudden end. This book explores certain questions that need to be asked about philosophy, science, faith, and religion. Let's start with philosophies at conflict with one another and at conflict with science, within a standard model of philosophy.

STANDARD MODEL OF PHILOSOPHY

What exactly is "philosophy?"

Let us use some high school algebra to understand this.

If x corresponds to the total ultimate truth and y corresponds to the portion of truth that is fully understood and established by science, then $z = x - y$ obviously corresponds to that portion of truth, which is either unknown or partly known and not yet established by science. The entire philosophy of the world, whether written in books/websites, or discussed in seminars/get-togethers, is dealing with z. That portion of z, which is related to such topics as mind, consciousness, soul, spirit, and God, is dealing with "religious philosophy." As and when a portion of z becomes an established truth by science (by experiment/observation), it is added to y. Until such time, it remains in the domain of philosophy/religion. The singular criterion to be adopted in arriving at one's religion is that it should be fully compatible with y and should not be required to reject an already established truth.

THE ELEMENTS OF THE STANDARD MODEL:

1) The laws of science must be such that the universe makes sense, and for the universe to make sense, there must be a consciousness to observe and to understand that the universe makes sense.

2) It is impossible to be aware that we are not aware, or to be not aware that we are aware. There may be time gaps between successive consciousness stages, but we are unconscious of these unconscious tenures, and they pass quickly.

3) An anthropomorphic—God, made up of atoms and molecules, occupying time and space—a miracle-performing, miracle-rewarding, and punishing God—may not be compatible with already established scientific truths "y." Thus,

such a God is impossible to believe. Equally impossible to believe is the concept of "No God at all," which would mean that there is nothing that controls nature and that nature acts on its own. This would lead to a universe that makes no sense at all. The only other alternative is that there is something out there, which controls nature in such a way as to ensure that the universe makes sense. It is not known what that something is that controls nature. It may take a long long time for science to understand what this something is. No harm will be done if we call this something God. And, no harm will be done if we assume the nature of this God as an "intelligent field" like an "all intelligent infinite mind." A component of this infinite mind (IM) encounters and interacts with a certain biochemistry that gives us consciousness and designs and builds bodies capable of internal resistance against external forces. Even such a component mind cannot on its own have control over these external forces, without violating y.

4) That there is a "law of causation," that every effect is preceded by a cause. The intelligent field ensures that even at the quantum level probabilities evolve deterministically via a certain mysterious connection with the classical world. (1-001). A serious understanding of the law of causation should reveal that even the Big Bang could not have been without a cause. (1-002)

5) That there is what we call the "anthropic principle," according to which the laws of science and the constants of nature are extremely fine tuned, in a way that ensures life and consciousness to emerge and then understand this universe. (1-003).

6) That there is a universal law of probability, which implies that there is nothing special about the human being in the eyes of the mind called God. This further implies that even though this may be a designed universe that ensures life

and consciousness to emerge, it could not have predicted when and where life and consciousness would emerge, or whether a civilization from a planet x could conquer another from planet y. Thus, if our human civilization is moving toward self-destruction—caused by the interactions of the universe, which in turn happen in accordance with the laws of causation and probability—the IM (infinite mind) may not come to our rescue. It is left to us to improve our interactions in a way that can enable us to create such probabilities that can save us from self-destruction.

These elements of the philosophical model, in particular the anthropic principle, are, of course, based on the initial condition that the universe makes sense. One might ask the question: Is making sense a requirement that the universe must fulfill?

Or is it as someone (a blogger Brandon Jett)said:

"We exist not in a universe of purpose, but one of absurdity and misunderstanding. I look beneath the sentient puddle that is humanity and watch the sun slowly evaporate us into nonsensical extinction..."

How does this description of the triviality of humanity and the universe measure up to the anthropic principle? Well, it's like this: The universal law of probability implies that while the universe may be intelligently designed for all kinds of lives to emerge in different parts of it, there is nothing special about human beings, nor is there anything special about any other kind of life in any other part of the universe. Again, the question may be asked: Why is so much of the universe inhospitable, and our planet so vulnerable to cosmological or geological events, as if to prove that it is perhaps not meant for us?

As far as inhospitality is concerned, consider that you are in a Supernova—an exploding star. There is nothing more inhospitable we can imagine, yet that is where we are manufactured. Supernovas are where all the heavier elements are created that become tables and chairs, trees and plants, birds and animals, you and me. And that is just one component/unit of this intelligently designed industrial universe.

The universe and humanity exist in spite of all odds piled up against. The designer is constrained. He needs to take the laws of causation, the second law of thermodynamics, and the law of probability into His calculations. This book in its entirety is an attempt to find the probable answers to the questions regarding the Universe, life, and consciousness. If our lives are ordered on the basis of these answers, we may be able to survive long enough to find the true answers to these questions.

This is an autobiography of a quark, from the Big Bang onward.

ENDNOTES

1-001:

Consider that I am walking along a road and need to turn left at the next turning to go to my destination. The quantum entities inside my body keep moving here and there "at random" with complete disregard to the phenomena of cause and effect. But, they will in their trillions be at the right places at the right times to ensure that I am turning left and not going straight or turning right.

In case something happens to me before I turn left, such as being hit by a car, the classical world (of the car hitting me) informs the quantum world of the changes in probabilities, and the quantum particles inside me will be at the right places at the right times to make sure that I fall down, provided, of course that the impact of the car is sufficient for the purpose. This is explained in great detail in chapter four.

1-002:

Does this mean that the Big Bang does not correspond to time zero? Common sense coupled with simple mathematics says that anything less than zero is negative. But is it possible to comprehend anything like negative time? The only solution seems to be that this region of time corresponds to the last stages of the previous universe or, more appropriately, the previous eon of this universe.

This means if we take the law of causation seriously that something occurred in the last stage of the previous eon, which happened to be the cause of which the effect was the Big Bang. Does this mean that this universe is not just an intelligently designed universe but a simulated one? Designed and simulated in accordance with the anthropic principle by "something," such as the superconsciousness of the previous eon. This subject is discussed in more detail in chapter six during "The seminar." However, this knowledge of the universe being a simulated one or not will not be available to us for a long time and is therefore not a part of the standard model of philosophy.

1-003

Some constants of the anthropic principle:

When we all arrived in the universe "as quarks" within the first microsecond after time zero, we had but one chance in several million (say x million) of surviving annihilation by our "antiquarks," and this ratio had to be very close to that value x in order that the universe would now be inhabitable by living creatures.

If the proton was just 1 percent heavier, it would decay into a neutron and all nuclei would become unstable and disintegrate. Atoms would fly apart making life impossible.

The ratio of force of gravitation and nuclear forces has to be precisely what they are in order for the wonderful stars to exist and for nuclear energy to continuously provide energy for our sun.

There are many constants of nature, which had to be in a certain exceedingly narrow range in order for life to exist in the universe. The subject is discussed at length in chapter six.

OTHER NOTES ;

BIG BANG: The discovery that the universe is expanding led to the realization that the universe was born in a Big Bang some fourteen billion years ago in a state of extreme high density and temperature both decreasing as the universe expands.

GALAXY: Large aggregation of billions of stars bound together gravitationally.

SUPERNOVA: A great stellar explosion when a star dies. (Refer to chapter two for more details.)

GLOBAL WARMING: Gradual increase in Earth's temperature caused by a buildup of carbon dioxide in the atmosphere (greenhouse effect) as a result of burning fossil fuels.

ANTHROPIC PRINCIPLE: The doctrine that the laws of physics as well as the constants of nature are specially designed in order to allow the existence of conscious life and thereafter to understand the universe.

Chapter Two

FROM TIME ZERO TO TIME NOW

I was there...during that first second...Perhaps even before that.
Yes I was there...in that quark soup...and so were you.

THIS IS MY AUTOBIOGRAPHY

I entered the universe as a quark. What time was it? It was 10^{-43} seconds after time zero, also known as Planck time "p." (2-001)

Something happened. Let's call it "the Big Bang."

A stupendous explosion *cum* inflation... How did it happen? Perhaps somebody switched on the computer, programmed by the cosmic consciousness of the previous universe.

There was radiation all around.

It was extremely hot, temperatures reaching one hundred billion degrees Kelvin. I entered the universe as a quark. (2-002)

I had but one chance in thirty million to survive. There were about thirty million quarks and the same number (minus one) of antiquarks in my neighborhood who annihilated each other in

9

no time at all. But one antiquark was missing, and I failed to find my counterpart and thus avoided annihilation. And so I arrived and so did you. Congrats! And so did all matter that makes up this universe, which is made of, mostly, this slight surplus set of quarks. Thus, I lost sixty million of my friends in no time at all — what to do? Well, it was programmed that way. If the antiquarks had been replaced by quarks, there would have been no annihilation, matter density would have been nearly sixty million times greater, and the expansion would have stopped abruptly causing instant crashing.

However, this large-scale annihilation of matter/antimatter resulted in a stupendous burst of energy in accordance with the equation $e = mc^2$ and this is what brought about all radiation. My freedom as a single quark was quite short lived. In less than a microsecond, two other quarks joined me. Together, as a trio of quarks, I became a proton or a neutron, the former having an 80 percent probability and the latter about 20 percent.

By the time I was about one second old, the temperature dropped to about ten billion degrees Kelvin and the density dropped to about two hundred thousand times the density of water. At this stage protons and neutrons started fusing into each other. There was a 25 percent probability that a few of these neutrons joined me, and I became an atomic nucleus, probably a helium nucleus. This fusion process went on for about three-and-a-half minutes and stopped abruptly as the temperature dropped to about 1 billion degrees Kelvin and the density dropped to about twenty times that of water. Most of the helium that exists today was forged during this brief period. A substantial portion of the universe (about 75 percent) did not participate in this fusion process and remained unprocessed, which means there is a very good chance that I remained as a single-proton hydrogen nuclei.

Next came about three hundred thousand years of relative peace. We, the hydrogen and helium nuclei, along with free electrons and photons (radiation) interacted in thermal equilibrium. The temperature, though reducing constantly, was still extremely high, the light still strong and bright, the radiation still energetic enough to separate electrons from nuclei. Expansion and cooling continued...

All of a sudden an abrupt change occurred. The temperature had dropped to about three thousand degrees Kelvin. At this temperature the radiation was not strong or energetic enough to prevent free electrons from combining with nuclei and decoupled itself completely from matter and proceeded on its way through space at the speed of light losing temperature all the while. That radiation has traveled for nearly fourteen billion years and is still moving. It is all around us and fills up all the intergalactic space. It is called "cosmic background radiation," and the current temperature is just 2.73 degrees Kelvin. And so, electrons could no longer be prevented from combining with nuclei and becoming stable atoms. Accordingly, I became an atom, and the odds, according to Ladbrokes, that I was a hydrogen atom are one hundred to twenty-five on and the corresponding odds for a helium atom as four to one against. In the language of quantum physics, you might say that I was in a state of superposition with an 80 percent probability of being a hydrogen atom and a 20 percent probability of being a helium atom.

Now, this radiation as well as matter had ample time to communicate with their counterparts in different regions of the universe prior to the decoupling stage. This aspect and considering that inflation earlier on had resulted in a stupendous zone of "causal contact" accounts for the fact that the universe is homogenous. Every entity, be it photons (radiation) or atoms knew exactly how they were supposed to act or behave in the future and could not act or behave in any other way.

Thus, at age three hundred thousand years, I, an atom along with others, entered the next phase of my life knowing precisely what I had to do. The laws of physics were spelt out to me in no uncertain terms by the anthropic principle, which in turn was designed and programmed by the cosmic consciousness of the previous universe.

The first thing I did with friends in my neighborhood (while still moving with the expanding universe) was to collapse into something bigger, which in turn collapsed into something even bigger (heavier may be a more appropriate term), which in turn collapsed into something even bigger, and so on. Call it gravitation if you like. This went on for millions and millions of years and is still occurring.

Let me take a forward leap in time. Around 3.5 billion years after time zero, I found myself to be a member of a certain star in a certain galaxy called the Milky Way. It was a massive star, nearly twenty-five times as massive as our sun. We (me and my colleagues) had been living there for nearly 25 million years while burning ourselves one by one from hydrogen into helium. We were about to witness the last days of the star, leading to an event of stupendously catastrophic magnitude and mind-blowing significance in terms of the next stage of the anthropic principle. We reached a stage when the entire central core was converted to helium. My own status at the time would have been one of the following:

a) I was already a helium atom near the centre.

b) I converted from a hydrogen atom to a helium atom.

c) I was still a hydrogen atom, being part of the remnant hydrogen still unprocessed near the surface.

The next stage was the initiation of the process leading toward the catastrophic death of the star.

What exactly happened in the next stage?

Ignition of the Helium

Balance life of the star, about one million years.

Sharp rise in temperature in the already hot star. Sharp increase in the density. Contraction. Conversion into carbon and oxygen.

This goes on until the temperature reaches about one billion degrees. Density reaches about one hundred thousand tons/cubic meter. Most of the helium in the central core is converted.

What's my status? I'm either an oxygen atom in the central core, a carbon atom in the region surrounding the central oxygen core, a helium atom left unprocessed in the region surrounding the carbon region, or a hydrogen atom still unprocessed near the surface of the star.

What's next?

Burning of Carbon

Balance life of star, about ten thousand years.

Contraction stops temporarily…temperature stops rising temporarily…carbon burning starts…conversion into magnesium.

This goes on for nearly ten thousand years. When most of the carbon is burned, nucleosynthesis picks up speed.

Death is now very close

What's next?

Formation of Iron

Balance life of star, a few days.

Contraction resumes at a highly accelerated rate…temperature starts rising again stupendously…nucleosynthesis proceeds at a staggering rate.

Formation of heavier elements like Silicon, Sulpher, Nickel, Cobalt, and iron near the center.

The star takes the shape of an onion with several layers of progressively heavier elements toward the center. Ignition takes place at the base of each layer, converting the elements in that layer to heavier elements in the layer below.

Central core dominated by iron.

Balance life of the star, two minutes.

What's my status? An atom of iron with maximum probability, or of silicon, sulphur, oxygen, carbon, helium, or hydrogen, *et cetera,* with progressively reduced probabilities.

What next? Stupendous struggle against gravitation collapse…creation of neutrinos in phenomenally huge numbers (probably of the order of 10^{57})…escape of neutrinos through the stars into space, leaking energy into space thus blocking energy to support the star against collapse. What next?

The Ultimate Collapse

A massive iron core is formed. Considering iron as the most energetically favored element, its fusion into still heavier elements such as gold does not release energy, but absorbs energy. As such, it cannot squeeze energy from fusion, and so there is nothing to prevent its gravitational collapse. In no time, density increases to ten billion tons/cubic meter.

Balance life of star, one second.

In less than a second, density near the center increased tenfold to about one hundred billion tons/cubic meter. That's about fifty thousand billion human beings cramped in a normal-sized room.

But, I couldn't bear it any more. I guess it was too damn overcrowded. We had to revolt, and we did just that. We revolted and crashed out. What else was there to do?

It was an explosion, a super shock of a mind-boggling magnitude. To give you some ideas of the magnitude consider the following:

If "a" is an event (in terms of magnitude) comprising a grain of sand falling on the floor from a tabletop, and "b" is the magnitude of an event resulting in the complete destruction of the planet Earth due to a nuclear or asteroid attack *whatever, and "c" is the magnitude of an event like the above mentioned super shock,*

Then $a/b = b/c$. And, this is not an exaggeration. In fact, it is an understatement.

Or consider the equation $e = mc^2$ with m as equal to several solar masses, and c, of course, as the speed of light. Now, try to imagine the quantum of energy created in the explosion. And that, my dear friends, was a supernova.

What has been captured in the above description of a typical supernova is just a broad outline. For a complete story of what exactly happens in a massive star during its dying minutes, refer to some of the books mentioned later. The phenomenon itself is wide ranging in its character and scope depending on, among other things, the star's mass (in terms of number of solar masses) with varying end results. The star may lose most of its material into space but retain some of it in the heavily contracted central mass and become a neutron star or even a black hole (a victory for gravity) or disappearing altogether, losing all its material by expulsion into space (a victory for entropy).

Was this a performance to empty stalls? No, it wasn't. It would be watched by intelligent beings of the future in far-off worlds.

Was this an industry? Yes, it was an industry to manufacture the elements destined to become the raw materials for future rocks, trees, buildings, tables, chairs, books, birds and animals, you and me. Nowhere else will you find them, except as products of a supernova. It was the next stage of the "Anthropic Principle," and I became an atom of carbon.

Subjected to the stupendous shock, I was at first dismembered to my original state (as in my birth second) of a proton made up of three quarks and later joined by other friends: protons, neutrons, and electrons with their own but somewhat similar past history. I then proceeded on my journey, with absolutely no idea where I was headed to

Suddenly I noticed something…I felt different…What was it?

"At the precise moment — in the aftermath of the Supernova — that I became an atom of carbon, a certain quantum awareness of 3.5 billion years crept in and became an integral part of me (culminating in the Supernova). But, it was just an awareness of a somewhat different kind. Coupled with this awareness was a property comprising a special characteristic, the desire/tendency to grow and bond with other elements. This desire will someday get fulfilled…God knows when…I didn't know at the time the extent to which the odds were stacked against me, that I must wait millions (or perhaps billions) of years before I attract and bond with other elements and become a molecular unit comprised of carbon, helium, oxygen, nitrogen, and other friends…in a precise mix…under a precise environment…at a precise temperature…as fixed by the designers. Once formed, this molecular unit would have the following characteristic features

1) It would be indestructible.

2) Its awareness quantum would increase by leaps and bounds over that of carbon alone…and it would be supremely intelligent as per the designed program, but not conscious in the same way as in living beings.

3) It would forever be on the lookout for a specific environment with a specific biochemistry to obtain a different kind of consciousness.

I will come back to this philosophical aspect later. At this stage, let me explain what happened to me and what, in my case that specific environment that I encountered was, what that specific biochemistry was that I interacted with, and what precisely the outcome of such interactions was: I take another forward leap and come to time "t," about 10.5 billion years after time zero when I arrived on planet Earth in search of that biochemistry.

"When I landed on Earth, perhaps three billion years ago, the planet was still boiling. Some of my companions fell deep beneath the surface, but I remained closer to the top. I was sometimes part of a landmass, sometimes part of an ocean sea floor, sometimes part of a swamp. And this continued for millions and millions of years. Most of my colleagues in my neighborhood remained inert. But something peculiar happened to me. I could sense that an electric field was leaking from the charged electrons and nuclei inside me. The stretching electric fields from the electrons inside me made me writhe and twist and pull myself into strange configurations. This odd phenomenon happened to many of my colleagues in my neighborhood but most of them fell apart. But not me…I was made of sterner stuff. I was determined not to let go this golden opportunity. I had waited more than ten billion years for this, and there was no way I was going to miss out on this. I looked up at the sun, absorbed the heat energy, and allowed the electric charges to take control over me and further contort the atoms inside me and allow myself to writhe and twist and pull myself to another configuration. In short, I had duplicated myself…and then quadruplicated…and so on."

The rest is history…

And so, …I take another forward leap…(2-006). And, I come to my present state of consciousness. At present I am residing in the cerebral cortex reticular formation, hippocampus, or medulla of the brain of a man called "SKS," dictating this manuscript and directing the brain to coordinate the flow of information and causing the fingers of his hand to type on the laptop. I travelled for nearly fourteen billion years covering nearly ten billion trillion miles before realizing I had traveled so much.

It has taken that much time (fourteen billion years or so) to reach this state of consciousness where I've had the privilege to develop an interest in cosmology, to be alive at the time when this branch of science has progressed by leaps and bounds, and to have interacted (through books of course) with the great scientists and cosmologists of the past and present. Now, it's another matter that I may not have been consciously aware of my existence all these fourteen billion years, and it's quite possible, and even likely that there might have been long gaps of perhaps millions of years between successive consciousness states. Even so, it can be said that I have been forever conscious for the simple reason that I was unconscious of my unconscious tenure, which period must have passed instantaneously for me despite being millions of years.

Regardless of the form of existence or the extent of awareness or consciousness, my journey from time "p" to the present through stars and supernovae has been intelligently designed so as to evolve from a freely moving quark (2-002) that obeys the laws of quantum physics but not the laws of classical physics. However, I am still participating in the evolution of the universe as part of a collective, obeying the laws of quantum physics as well as those of classical physics. (2-003) Such evolution of the universe is also intelligently designed in such a way as to proceed toward creating an environment with an appropriate biochemistry ideally suitable to gradually improve the degree of awareness from an infinitesimally low level to a sufficiently high level of consciousness where the universe can understand itself and write down its own resume. And this is what is called *the anthropic principle*.

The difference between science and religion is that, according to religion, it is God who created the universe. According to science, it is the anthropic principle that matters, and the same has been given the name of God in the English language.

At this stage, let us briefly try to look at the future as an extension of the same anthropic principle, according to which the degree of consciousness with special reference to the understanding of the universe is forever increasing. The question is this: "Can this degree of consciousness at some time in the future reach the stage of cosmic consciousness? Not only is this possible, it is also probable. It might appear that fourteen billion years is a long time to achieve the current level of consciousness.But if we look at the current understanding of cosmology, it is likely that the life span of the universe may run into trillions and trillions of years, in which case fourteen billion years is a very short time, the universe can be considered to be in its infancy, and there is ample time for full-fledged cosmic consciousness to evolve. A bookmaker blessed with a full understanding of cosmology would probably give the following odds (hypothetical of course) for the probable life span of the universe:

1) For life span exceeding 10^{100} years: ten to four in favor.

2) For life span between 10^{20} and 10^{100} years: three to one against.

3) For life span between 10^{11} and 10^{20} years: thirty to one against.

4) For life span of the universe Less than 10^{11} years: two hundred to one against

The first three correspond to an open universe, which keeps expanding forever, and the last one corresponds to a closed universe, which could expand for another twenty to thirty billion years before gravity takes control and the universe starts collapsing toward the Big Crunch. This last case is the least likely of all possible scenarios as the energy density is too small for gravity to develop sufficiently to be able to stop the expansion.

However, in the last case (lifespan less than one hundred billion years) cosmic consciousness could still develop considering that the relation between degree of consciousness and time elapsed after the Big Bang has not been a linear relationship and that the rate of increase in the degree of consciousness per unit of time itself has been increasing nonuniformly with occasional massive spurts of increase.

But, if the universe proceeds on expected lines, expanding forever, it may, after the present, exciting Stelliferrous Era — where stars are forming continuously, with many of them supporting planets with life on them — pass through enormously prolonged dull and uninteresting eras such as the Degenerate Era, the Blackhole Era," and the Dark Era. (2-004)

What happens in the universe during these eras is enormously complicated and filled with intrigue and uncertainties. There is no reason to believe that life may not exist during these stages. But if it exists, it may be in a completely different form, impossible to imagine at this stage. If there is no life during these periods then the "Anthropic Principle" itself must be flawed, for there is no point in having life for less than one trillionth of a trillionth of the universe's life. What then is the scope of the Anthropic Principle? Will life in these periods of the future, in whatever shape it is going to take, be the ultimate form in which it can exist? Or, is it that cosmic consciousness is destined to emerge long before that? Could it possibly play a part in the future history of the universe by such acts as changing the laws of nature (or physics), such as reversing the direction of entropy — the second law of Thermodynamics — or by some other course of action. We are free to imagine what we would like to imagine. It has been said that this universe is indeed a queer universe; it is not only queerer than what we imagine, it is queerer than what we can imagine. Can it be that this cosmic consciousness, after some considerable period of running the universe, relaxes for some time in an energy-free environment? This period of relaxation may be called *"pralaya."* (2-005) Could it be that during this period of *pralaya,* this cosmic consciousness in silent meditation prepares the groundwork for the next universe? Does it study and analyze the past, what went wrong and what went right, and prepare a comprehensive program?

An important issue to be resolved here is "How does consciousness creep into the carbon atom." Or, in other words, "Who is this I?" or "What is this substance called I?" Is this matter? Was the quark I, who arrived in this universe at Planck time after time zero, one of the constituent parts of this substance I, which is currently residing within me?

But, it is said that each and every atom inside the body is replaced every five to seven years. Could it be that the "quark I" of the Big Bang that entered "me" at the "speck of jelly" stage has left me and is no more inside me? How can that be a possibility? The information that has been passed on to me by the influence of the world for the last seven decades, ever since that "speck of jelly" stage, onward is still inside me. It is either entirely contained in a certain substance called I or surrounding it all the time in a narrow region of the brain. Before these atoms are replaced the information gets transferred to the new incumbents and thus forever remains inside me. What then can be that substance called I?

Now, I am going to speculate with another "wild imagination." The substance called I is none other than that "molecular unit" referred to earlier. Apart from those characteristic features mentioned, it is forever on the lookout for a specific environment with a specific biochemistry. It attains fruition with the substantial interaction of something called "M," a sudden upsurge in the level or degree of awareness/consciousness. It immediately starts accumulating matter around itself, gets contained inside it, and together with its container becomes life and in the course of time will be called so and so.

It's all intelligence, and its intelligence quantum is commensurate with the sum total of interactions and experiences obtained during billions of years inclusive of those of its component parts before they all came together. Commensurate with the intelligence available to it, it will design and execute the growth of its container with optimum precision; it will have no control over the external interactions of the rest of the world acting on it. It will, however, absorb those interactions that are beneficial to its growth and build internal resistance against those interactions, which are harmful to it. Broad parameters of growth shall be: painless existence of the container for an optimum duration of time, a further increase in knowledge (and thus intelligence level), and understanding of the universe for itself.

Let's call it "The Mind, Body, and Soul" with the molecular unit as "soul," the container as "body," and that something called M as "mind." The molecular unit along with that something called "mind" together constitute the "self" of the container and can be considered as its personal God. All of this was programmed by the "superconsciousness" of the previous cycle of the universe. However that superconsciousness had no control over the law of probability. As that super intellect switched on the computer and set the ball rolling for a new universe, it was understood that it would have no control whatsoever on when and where lives would emerge or whether a civilization from planet x will conquer another from planet y or whether Roger Federer would win the next United States Open.

And who knows, when we are a trillion times as old as we are now, you and I and a "scillion" other molecular units will be part of that superconsciousness that will design the next universe. The erstwhile designers would have taken care — in their calculations — to make sure that we will be provided with the appropriate biochemistry — or maybe biophysics — to remain in that conscious state while we are at work until the job is done.

ENDNOTES

2-001:

Planck time: The time taken to cover Planck length while moving at speed of light. Planck length (about 1.6×10^{-35} meters) can be considered as the smallest region of space that can be described meaningfully, and can be arrived at by combining the constant of gravity, Planck's constant and speed of light. For Planck's Constant refer to chapter four.

2-002:

quark: "Five distinct levels of matter have been identified: the molecules, atoms, nuclei, hadrons, and quarks. Eight dozen atoms were a simplification over their millions of molecular compounds. The nuclei of eight dozen atoms were bound states of just two hadrons, the proton and the neutron. Each hadron could be viewed as made up of few quarks orbiting around each other in specific configurations. quarks are point quantum particles similar to the electron with the same spin of one-half but only a fractional electric charge compared to electron's unit of charge. No one has seen a quark. The quark model was in fact invented to simplify (mathematics) the complexities of the hadrons. The quarks appear to be point particles, without further structure—a "rock bottom" of matter. In the future additional levels may appear, but so far there is no evidence of the same." Extracted from *The Cosmic Code* by H. Pagels. (Refer Part two of the book, `THE VOYAGE INTO MATTER Chapter 1 `The Matter Microscopes` page 181 and chapter 4 `Quarks` page 198 etc)

2-003:

Quantum physics and classical physics: Quantum physics is the science of the very very small—say of size x and below—where Newton's laws are not applicable, as they are in classical physics. There is no clear-cut demarcation of the size x, as it is a measure of the quantum of accuracy required in the measurement process.

2-004:

"Degenerate Era: Dead stellar remnants capture dark matter, collide with each other, scatter into space and finally decay into nothingness.

Blackhole Era: Black holes inherit the universe, warp space and time, evaporate their mass energy, and make an explosive exit.

Dark Era: The nearly moribund universe struggles with cosmological heat death and faces the possibility of universally transforming phase transitions. (Extracted from *The Five Ages of the Universe* by Fred Adams and Greg Laughlin.)

2-005:

Pralaya: All space will be at the same temperature. No energy can be used because all of it will be uniformly distributed through the cosmos. There will be no light, no life, no warmth—nothing but perpetual and irrevocable stagnation. Time itself will come to an end, for entropy is the measure of randomness. When all system and order in the universe have vanished, when randomness is at its maximum, and entropy cannot be increased, when there is no longer any sequence of cause and effect, in short when the universe has run down, there will be no direction to time, there will be no time. And there is no way of avoiding this destiny. Extracted from `Modern Physics and Vedanta` by Swami Jatatminanda

2-006:

This last forward leap pertains to the period pertaining to the journey till my arrival sometime in the mid-twentieth century and immediate containment in a speck of jelly. The said container was to be called SKS. Some understanding will also be required of a certain phenomena called "Morphic Resonance." Men like Charles Darwin (*Origin of Species*), Erwin Schrödinger (*What is Life?*), Rupert Sheldrake (*Hypothesis of Morphic Resonance*), and John Gribbin again (Stardust) and a few others, including Stephen Jay Gould, will hold center stage in that section.

However, there is something else that must be understood in order to understand the above phenomena of that biochemistry and "M.R." and that something is called quantum physics (In as much as "Life itself is a Quantum process").

If you are feeling that your mind has been boggled to some extent on reading about this supernova business and what happened after that, let me tell you that compared to quantum physics, it is just a "T.O.T.I." (Tip of the iceberg)But we must wait awhile. Quantum physics will be the subject of chapter four. Before that...a little peep into the present century on our beloved planet Earth, an exciting century no doubt, but a difficult one to travel and survive through to the next. And the next...Unless!

For further reading:

1) *Cosmology* by Edward Harrison

2) *Stardust* by John Gribben

3) *The Five Ages of the Universe* by Fred Adams and Greg Laughlin

4) *What Is Life and Mind and Matter* by Erwin Schrödinger

5) *Cosmic Code* by H. Pagels

6) *Electric Universe* by David Bodanis

7) *Modern Physics and Vedanta* by Swami Jatatminanda

Chapter Three

EXCITING CENTURY HOW MANY MORE CAN WE HAVE?

This is an exciting century…But it could be our last…Or last but one…

Consider the statistics, arrived at after a detailed study of the population records of the planet. In one hundred thousand years (ending 1900 AD) the population of the world increased from a few thousand to about two billion, in the next one hundred years it trebled to six billion; presently it stands at 6.9 billion. Ten percent of all the human beings who ever lived on the planet Earth are still alive today in the year 2014 (assuming about 62 billion deaths so far).

If the figure "10 percent" in the above statement is considered as the P Factor corresponding to the year 2014 — this factor keeps rising with time — its value was only about "3 percent" in the year 1900. The rate of increase of population has also been increasing though nonuniformly, from trebling in five thousand years at one stage near the beginning to trebling in one hundred years in the last century. Even if we assume there is no further increase in the rate of increase, trebling every one hundred years, the factor P will increase to about 21.5 in the year 2114, and to about 36 in the year 2214. Thus in the year 2214, we could make the following statement:

Thirty-six percent of all human beings who have ever lived on the planet Earth are still alive today in the year 2214 (assuming about sixty billion still living and 104 billion deaths so far). From the above, it looks like that not only all who are leaving are coming back again but there are others who are joining us on a regular basis, perhaps from other life forms.

Perhaps it may not be incorrect to assume that "not a single human being who was ever at any stage alive on this planet Earth is not alive today." But these population levels will not be sustained. Wisdom will have to eventually kick in and eliminate the folly of overpopulation. But a serious realization to that effect has not manifested itself till now. Nothing much is happening in this regard. If the rising trend continues, nuclear wars will become imminent.

The notion that large-scale conventional wars may help in reducing population levels is not correct. The twentieth century witnessed several wars including two World Wars. The number of lives lost was unmatched in any previous century, yet the real population explosion took place in the twentieth century. In fact, the rate of increase of population experienced a sudden surge around the year 1930. Nuclear wars followed by chain reactions are the only way of causing sudden massive falls in world population, but is also the only way of causing the extinction of the human race from the planet.

Even If we assume that the probable chance of a nuclear attack is as low as 1 in 200 per calendar year — and a probable chance of the conflict going global as 1 in 4 for each nuclear attack — the probability of self-destruction can be worked out as nearly 50 percent in two-hundred-fifty years and in excess of 90 percent in the next five hundred years. This is a chilling thought. And time is running out for us. It is high time that the "survival of human race" must become the most important subject to be taught in universities all over the world.

And the syllabus must include convergence of science and religion to end conflicts between nations, not to mention within nations, and the best bet to ensure such a convergence is an understanding of reality that has the best chance of acceptance by all religions. By far the biggest obstacle that is coming in the way of a convergence between science and religion, not to mention convergence between different religions is an improper understanding of the word "God."

It is imperative that we must first arrive at the correct definition of God before talking about God's existence or nonexistence. If we go by the dictionary meaning of God as a superhuman being who controls nature, then it can be said, by applying the laws of science that such a God cannot and does not exist, and thus we are all atheists. On the other hand if we define God as something that controls nature, but we do not (or science does not) *yet* understand what that something is, then by dint of this definition alone, it is proved that God exists, and thus we are all theists.

There is a third alternative that there can be "nothing" that controls nature, which would mean that nature acts on its own. This last alternative can preferably and "for all practical purposes" be completely ruled out by consideration of the following:

1) It can never ever be proved.

2) It cannot explain the Big Bang

3) It cannot explain the phase transition that brings entropy from a near maximum to a near zero required at the Big Bang.

4) It cannot explain the selection of the constants of nature.

5) And above all, even if there is a remote chance (0.00000001 percent) that it might be true, it leads to a universe that makes no sense at all. (3-001)

This is as simple as it gets and should freeze all discussion on the subject pertaining to God's existence or otherwise.

Now, having come to an understanding that there is neither an anthropomorphic, physical God, nor a nothing-type God, and that there can only be a something-type God, let us look at the second hemisphere of our brain (3-002) — the one on the right side. Let's use our intuition and try to guess what that something could be, which has the best chance of its acceptance by all religions, and a reasonably good chance of being true.

Here is a possible answer: "An all intelligent omnipresent — in space as well as in time — infinite mind." And, it's the same mind everywhere in each of us. But, for the specific I in the physical me, which includes the brain, the world in its particular form is given to me only once. As Schrödinger said: "This world extended in space and time is but my representation, experience does not give me the slightest clue of its being anything besides that." Extracted from ``What is Life`` by Erwin Schroedinger

Now there can be two alternatives: That there is a mind and in addition there is a soul. That the I in me is a soul. It has a permanent identity; it is none other than that "molecular unit" referred to earlier.(3-003) The mind is all in all…the singular as well as the plural…the part as well as the whole. Each alternative ensures immortality. But — in both alternatives — the conscious mind, which is "in everyone, when it has entangled itself with the brain of the physical me, is only looking after its owner. In my case "me." With all this, there is no doubt that the most important question remains unanswered and that question is: How is it that events in time and space that take place inside a living organism are unaccounted for by the normal laws of physics and chemistry? Why is it that these events are different when they happen outside the living organism and easily explained by physics and chemistry?

For example, how is it that even a single-celled creature — but still a living organism — such as a paramecium, can swim toward food, retreat from danger, and negotiate obstacles — thus disobeying Newton's laws of motion — whereas a dust particle — outside a living organism — can do nothing of the sort?

People might say "that's nature" and be done with it or that we should leave it to the theologian to answer the question. But we need to understand the science of it. Where did this intelligence come from? This intelligence in the form of an "intelligent field" pervades all space. The quantum of this intelligence can be gauged from the fact that a single-celled paramecium is intelligent enough to swim toward food, retreat from danger, and negotiate obstacles, *etcetera*.

Now it is obvious that present-day science is unable to answer these questions, but that does not mean science will not find the true answers in the future, maybe the deep future.

Yes, There will be ample time to understand these things. But a few things need to be understood quickly: That this mind is in pain, with too many lives, too many creatures racking each other in everlasting strife, too much oil spilled on the seas, too many conflicts and wars, nuclear arsenals lying handy, too great a probability of extinction of life on the planet.

In this universe of ours, no one is paying attention. No matter, it's just our representation of the world that is dying; the mind will find another abode, another biochemistry, on another planet, on another star. But in a parallel universe, these issues are taken seriously. Survival of the human race is the most important subject of studies worldwide.

And I am moving on (Remember Hank Snow?)...

And taking you along with me to that parallel universe...

Where, too, this is an exciting century...

But there is a chance, we will survive the collapse...

But before that...

We must understand "The Quantum." (3-004)

ENDNOTES

3-001:

The universe must make sense: a) If the universe began without the specified initial conditions, it would have to be one in several billion universes in order to have a good chance of having the desired constants, if it were to be like our universe brimming with life and consciousness. That's enormously huge metaphysical baggage to be carried, an extremely uneconomical universe that makes no sense at all.

b) Even in this one in a billion or so universes, life and consciousness can exist for not more than about 10^{18} years (since the Big Bang), i.e., in the Stelliferrous era where galaxies and stars are existing. Thereafter, the ensuing degenerate/black hole/dark eras, where dark energy keeps driving the universe toward perpetual nothingness, provide us with no possibility of life and consciousness. And, these eras go on endlessly to perhaps 10^{100+} years (stbb), which means there is life for only one unit of time out of nearly 10^{82} units of time.

See chapter six for more details.

3-002:

As Amaury De Reincourt said: "It is interesting to note that neurophysiology now tells us that the two hemispheres of the brain each specialize in different functions of cognition – the left dedicating itself to logical and analytical thinking, seat of verbal thought; the right to intuition and holistic understanding of patterns, with an ability to grasp directly the relationship between parts of the whole. One operates linguistically with rational sequences of deductions and inductions; the other intuitively juxtaposes images and symbols, integrates and synthesizes rather than analysis. It seems obvious that Western philosophic and scientific culture has, so to speak, atrophied the right hemisphere of the brain, while Eastern culture has given it predominance." (Extracted from *The Eye of Shiva* by Amaury De Reincourt).

For more detail refer chapter six.

3-003:

Refer to chapter two.

3-004:

Go to chapter four.

Chapter Four

QUANTUM ENTANGLEMENT: I AM ALL SEWED UP... ENTANGLED...IN THIS "NONLOCALITY"

What exactly is this nonlocality? Before answering this question let me ask another: What is locality or local causality? A hits B and B falls down. A predator attacks and kills its prey and starts eating it. The various component parts of the prey's body start moving toward the inside of the predator's body. A batsman hits a cricket ball. The ball, which was hitherto traveling at a certain speed in a certain direction, changes its direction and speed. If the ball then travels toward a certain line and crosses that line, the commentator calls it a boundary, and the scorer then adds four runs to the batsman's as well as the team's score (also think of baseball). On the other hand, if the ball goes high in the air and is caught by a fielder before it falls to the ground, the commentator says "He's out!" In a few moments TV News channels across the world display *"Breaking News—*Another Wicket Falls." Millions of people in different parts of the world come to know of it and experience a sudden increase in the coefficient of happiness or disappointment, depending upon which team the batsman belongs to.

It's all local causality or simply cause and effect. Everything that you can think of that happens in this universe—mind you, at the classical level only and not at the quantum level—is nothing but cause and effect; all actions lead to reactions and interactions, *et cetera*.

So far, so good—no big deal at all—it's fully understand-able—it follows Newton's laws of motion—no problem whatso-ever. Consider now the following: A hits B in Bangalore...C feels the pain in Chicago. Is this possible?

There are two categories of pain: mental pain and physical pain. The former is a distinct possibility and very well explained by local causality. C is a friend of both A and B and the informa-tion regarding the fight between A and B conveyed by phone to C causes mental anguish and pain to C. If this leads to mental stress that causes severe headache or pain in the chest, we can infer that physical pain to C is also attributed to local causality. But what about physical pain, instantaneously, without any in-tervening phone calls? That sounds impossible. At the classical or macroscopic level it does not happen.

However, at the quantum level it happens all the time. (4-001) Yes. This does happen in the universe, and this is a feature of quantum mechanics. And, this is what we call "nonlocality" or "quantum entanglement." How did the notion that there can be something as mind-bogglingly difficult to comprehend as non-locality become conceived in the first place, and who could have thought of it? It defies common sense completely. Indeed, it is as difficult or almost as difficult to believe as the statement "The Earth is flat."

Once again it's the great Einstein who comes into the picture as the sole originator of this grand idea. Einstein, who was one of the few great scientists of the world responsible for the emer-gence of the science of quantum physics, was always of the opin-ion that quantum theory, despite its huge success, was not yet the real thing—that it was somehow incomplete that there must be certain "hidden variables," which are not known to us now, but in time to come will reveal themselves to us to complete the picture.

Einstein, together with Podolsky and Rosen, came out with a stunning thought experiment called the EPR Experiment, or the EPR Paradox that demonstrated in no uncertain terms that if quantum theory is true then it leads toward the possibility of a nonlocal action. But, according to Einstein, as this nonlocal action (without causation), which he called "spooky action at a distance" was absurdly impossible, it must be understood that quantum mechanics is incomplete.

The EPR Paradox confounded scientists all over the world. Quantum physicists had no answer, though they were in no mood to reconcile to the conclusion that quantum mechanics was an incomplete theory. For nearly thirty years the issue remained unresolved.

Then came a scientist called Bell—John Bell with a profound idea—with a solution that it is possible to determine whether or not quantum physics is a complete theory or an incomplete one, that an experiment can be devised to understand whether nonlocality is indeed a feature of quantum physics. There is a certain expectation based on probability theory of a certain "inequality" (to be explained later) in the outcome of an experiment, which expectation is based on the normal common-sense view. In case the outcome of this experiment reveals a violation of this inequality, then it is proved conclusively that nonlocality or quantum entanglement is true. Later, experiments were actually carried out in practice. Certain experiments carried out by the French scientist Alain Aspect demonstrated clearly that the said inequality was violated thus showing that quantum mechanics was true. There was no incompleteness about it; at least there was no proof in favor of incompleteness, and if it led to the assumption of non-locality via quantum entanglement, then so be it.

Scientists and quantum physicists of today around the world are _not_ in disagreement on *what* happens at the quantum levels. They are only in disagreement on *how* it happens. And that's why there are several interpretations of quantum physics such as The Copenhagen Interpretation, The Many Worlds Interpretation, The Transactional Interpretation, and String Theory, to name a few frontrunners, to be discussed in some details in this and in subsequent chapters.

I cannot hope to get away without explaining the EPR Paradox, Bell's theorem, and what is meant by Bell's Inequality, and how Aspect carried out his experiments to prove the truth regarding nonlocality. I have a certain favorite set of books on the subjects referred to above, which I strongly recommend that the reader should read, particularly in case the reader desires to catapult himself into the ultraprivileged category. (4-002) Some of these are now lying in front of me, with the relevant pages open on my study table on this Saturday morning after breakfast, with an expected two rounds of coffee. Rachmaninov's Piano Concerto No. 3, playing on the music system. Let me see how best I can assimilate the picture and transfer it to my brain and what best I can do to compose the rest of this write-up on this queer phenomenom called quantum entanglement. However, before I get to this EPR thing, I need to dwell a bit on what went on in the quantum world, which led to the EPR Experiment — in short a brief history of quantum physics.

I tell you, I am in for some fun. And as for you…I hope your seat belt is fastened.

1900, Max Plank — Blackbody Radiation; The Quanta

Consider a metal bar — a black one — inside an extremely dark room, so dark that the black metal bar cannot be seen. Now take it out of the room and heat it on a fire to a very high temperature. Bring it back into the dark room; it's no longer black. It's color has changed to red and it's glowing and visible. Take it out and heat it to a still higher temperature and bring it back into the dark room; the color has changed to orange. Repeat the process. The color changes to white.

What exactly is happening and what precisely is the physical basis of this phenomenom?

This change of color and the distribution of this change are measurable and recorded as a black body radiation curve. This curve is a relation between the wavelength of radiation and the level of energy. In fact, there is a formula linking this wavelength with the temperature called the Rayleigh-Jeans law. But this formula is based on classical physics, though it gave results, which tallied with the experimentally observed intensities of radiation up to a certain frequency levels. It was found to be inaccurate for very high frequency levels, showing intensities of radiation, which did not tally with the observed intensities for a hot (black) body. It was expected that this energy that goes into the field from the black body would keep rising with the frequency and would reach catastrophic levels at extreme high levels of frequency. But this doesn't happen. At a certain specific frequency level (corresponding to color) for a given temperature, energy reaches a peak level and thereafter starts reducing.

Why was RJ's formula—classically calculated—showing discrepancies at high frequency levels? The scientist—Max Planck—studied this phenomenon extensively. In fact, he spent a considerable time on it. But just when he was about to give up, in sheer desperation he came up with a stunning idea that this flow of energy was not a continuous process. It came in packets of energy called "quanta." He opined that the atoms of the black body when heated to higher temperatures became vibrating oscillators, and these magnetic oscillators can occur only in discrete units. In short the energy exchange of these atoms with the black body radiation was quantized. The extent of this discreteness, or the value of this "quanta" was determined and specified by Planck as a constant "h," known as Planck's constant The value of this constant was worked out as 6.6×10^{-34} Joules per second. It is tiny but not zero, the difference being just about enough to differentiate between continuity and discontinuity. With the help of this hypothesis, Planck arrived at a relationship between energy and frequency that fitted remarkably well with the experimentally observed values.

Indeed, this was an unprecedented leap of imagination, *and quantum mechanics was born.*

1905, Albert Einsten—The Photoelectric Effect; The Light Quanta; The Photon

Notwithstanding the leap of imagination, Planck's revelation did not get much attention at the time; its significance was not understood so well, until Einstein came out with another staggering proposal—that light, which is nothing but electromagnetic radiation, must itself be made up of discrete units. Einstein used Planck's hypothesis to explain the photoelectric effect. If a beam of light strikes a metal surface a small electric current is created. Solar cells convert ordinary sunlight into electric power, automatic elevator doors open and close, TV Cameras work, and so on. Einstein received the Nobel Prize for the discovery of this photoelectric effect (Note that he did not get a Nobel Prize for his special relativity theory.) Einstein reasoned that if the flow of energy was in discrete units, then light itself must be made up of discrete units—the light quanta. Later this light quanta was christened "photon." What does this mean? Is light after all made up of particles?

But light was already proved to be a wave (Refer to Thomas Young's double slit experiment"), even though it was initially considered as made up of particles or corpuscles in Newton's time. What exactly is the true nature of light—particle or wave?

The immensely intriguing story of the understanding of the phenomena of light is best described in the book *In Search of Schrödinger's Cat* by John Gribben. Light was initially considered and shown to be made up of particles by Newton. Several scientists such as Christian Huygens (a contemporary of Newton), Leonard Euler (eighteenth century mathematics and inventor of pi), and Benjamin Franklin were certain that light is a wave. This was confirmed by Thomas Young nearly one hundred years later with his experiment. But such was the impact of Newton's personality that scientists were not ready to accept any ideas that opposed Newton's. Nevertheless, support of the wave concept continued and gathered fresh momentum with the experiments of French scientists Augustine Fresnel and Leon Foucault (of the pendulum fame). Then came the Scottish physicist James Clark

Maxwell and the German scientist Hertz. The former predicted radio waves and the latter confirmed these with his experiments) and the wave idea was fully established. By the end of the nineteenth century, the picture was completely reversed from particle to wave. So complete was this reversal that only a genius or a complete fool could question its authenticity. That gentleman was none other than Einstein (albeit with some help from Max Planck).

That leaves us with the unanswered question: Is light a particle or a wave? This was answered by Einstein himself. In a lecture he gave in the year 1909, Einstein showed the dual nature of light that it can simultaneously have both particle- and wave-like properties. This idea was supported fully by experiments, such as the double slit experiment.

1911, Ernest Rutherford — Atom as a Mini Solar System

The structure of the atom was understood for the first time in the year 1911 by Rutherford. It consisted of an extremely tiny, positively charged core called the nucleus with negatively charged electrons orbiting around it in the same way as planets orbit around the sun, except that instead of gravity, electric forces bind the system. The nucleus is so tiny compared to the atom; it is like a grain of sand in a huge auditorium. But virtually the entire mass of the atom is in the nucleus with the electrons' mass quite insignificant in comparison.

1920s: Niels Bohr, Werner Heisenberg, Louis de Broglie, Erwin Schrödinger, Paul Dirac — Mechanics of Particles and Waves

For the next few years after Einstein's 1909 lecture, nothing much happened in the development of quantum physics. In this period, Einstein's special theory of relativity as well as his general theory of relativity took center stage.

The next big leap of imagination was by Niels Bohr. The stability of the atom (why the electrons were not collapsing into the nucleus) could not be understood; it could not be explained by classical physics. Bohr considered Planck's and Einstein's ideas and realized that a new kind of physics must emerge to explain the various phenomena at the atomic level. The fact that the energy was quantized showed that electrons could jump from an outer orbit to an inner orbit, but only to a limited extent to ensure the stability of the atom. The energy lost by the electron while jumping from the outer to the inner orbit was emitted as light by the atom. Since the electron orbits were prespecified in view of the quantization of energy, the light that was emitted was also quantized. Based on the above model, Bohr calculated the light spectrum of the hydrogen atom and the results tallied with the experimental observations.

The Twenties must hold the record for the maximum number of "giant leaps of imagination" per decade. The next big leap of imagination came from Werner Heisenberg. He was struggling with the problem of understanding the atomic spectral lines. He was also suffering from hay fever and decided to go to an island to recuperate and clear his head. With no distractions, he was utterly focused, and in a flash of brilliance it occurred to him that he must approach the problem in a different way by considering not what the atoms are, but what they do. With this approach, the mind became crowded with thoughts full of intriguing possibilities. A stage was reached when sleep was out of the question. At three in the morning on a certain day at the said island, a massive leap of imagination did happen to Heisenberg. The picture of what these atoms do during energy transitions was revealed to Heisenberg in the form of an array of numbers placed in certain mathematical order and following certain rules of mathematics, which these numbers obeyed.

This leap of imagination–phenomenon in itself is a subject of immense intrigue. How does this happen to a brain, which is itself made up of electrons and protons? What happens to the atoms, which constitute the brain? What do these atoms do at these times?

While Heisenberg himself was a great mathematician, he was not aware that these rules of mathematics where already known as matrices. At the time he was not aware (but later learned quickly enough) that these matrices have a unique feature: they do not commute—the product of two matrices depends on the order of multiplication. In short, A x B is not equal to B x A. He returned from the island and revealed his ideas. A flurry of papers followed, beginning with his own, followed by a joint paper with Max Born (who knew of these matrices; Born was an assistant to Heisenberg) and Pascual Jordan (Born's student), known as the "Three-man paper." In this paper an equation was determined, linking quantum variables such as position and momentum as matrices, with Planck's constant "h" and "i" the square root of -1. This was followed by a paper by Pauli who was a master in the mathematics of matrices wherein he solved the problem of the light spectrum of the hydrogen atom and arrived at the same result as Bohr. Finally, it was the brilliant Paul Dirac—the twenty-three-year-old mathematician *cum* physicist—who wrote four papers covering the subject in a comprehensive manner formulating this new matrix mechanics. Incidentally, Paul Dirac was a qualified engineer who did not get a suitable position as an engineer and took to further studies in mathematics, which he took very seriously. His paper established quantum theory as a dynamic theory, which replaced classical mechanics. In fact, Classical mechanics could be derived from quantum mechanics by taking Planck's constant as zero. The mathematics was of course difficult to understand, but to those who understood it, the papers were doubtless sheer dramatic works of genius. However, the enormous difficulty faced by the scientists in understanding the mathematics of this theory was one of the reasons why these scientists were somewhat drawn away from it. The other reason was the emergence of an alternate theory called wave mechanics. Incidentally, Dirac also played his part in this and was instrumental in bringing wave mechanics to its final shape.

In a parallel sequence of events, in 1923, Louis de Broglie reasoned by analogy with Einstein's photon concept that if light, which is clearly a wave can also be a particle, then an electron, which is clearly a particle can sometimes be a wave. In two papers written in 1923 by de Broglie, this idea was presented emphatically. He even deduced the wavelength of the electrons. It was also verified experimentally that electrons showed diffraction phenomena attributed to waves. The mysterious nature of atoms could be explained by this concept that matter had these kinds of wavelike properties.

Erwin Schrödinger heard of Broglie's work and understood the significance of the idea. Another leap of imagination of an unprecedented magnitude followed. Schrödinger developed an equation whereby an electron wave shape would have to obey as if part of a hydrogen atom. The equation was of far-reaching significance. It reveals what happens inside an atom in a way that makes it possible to calculate the properties of more complex atoms and molecules. Paul Dirac even said that the entire science of chemistry can be derived from the Schrödinger equation. Dirac and Schrödinger found out that the matrix mechanics of Heisenberg and wave mechanics were completely equivalent and gave an accurate description of the quantum phenomena with the same end results. Wave mechanics was the more easily understood of the two and was therefore more popular with the scientists of the time and extremely successful as a practical tool. Dirac, in fact, combined the two theories into his "transformation theory."

Quantum theory was now complete with the mathematical formulation of the quantum phenomena very much in place. Two categories of scientists emerged in relation to quantum theory.

Category One — THE INDIFFERENT SCIENTIST

This includes the engineer, the chemist, the molecular biologist, all concerned with what happens at the quantum level but not much concerned with *how* it happens. The impact on technology and the practical applications of quantum mechanics were far reaching. The laser, the transistor, the semiconductor, the microchip, the computer revolution, DNA, RNA, and the biological revolution started changing the world. Chemistry was fully explained and nuclear physics was born. These category one scientists, though concerned only with *what* happens at the quantum level, had their own leaps of imaginations, one after the other, leading to these technological revolutions. Prior to the establishment of quantum theory, the word "computer" was used as a designation meaning "one who computes." These were people who worked on mechanical calculators. Quantum theory, revealing what goes on inside atoms and how electrons behave, gave specific insights into how electricity could be utilized for this purpose. The single most important aspect that led to these leaps of imagination was the quantum phenomenon that an electron can jump from one point "a" to another point "b" without having actually traveled between "a" and "b" in a continuous way. This is precisely *what* happens, and this was the "knowledge" that was utilized by hundreds of practical scientists of category one that brought about the electronic computer.

Another phenomenon was the peculiar behavior of electrons inside elements making some of them suitable as conductors and others as insulators. In the former case, such as in steel or copper, the electrons, which are in the outermost orbits of their respective atoms, tend to get attracted by intruders—such as photons of light or electrons from another source—and move along with these intruders as electric current. In the latter case, such as glass or wood, they are happy and contented to remain within their respective atoms. They are not in the least bothered by these intruders and remain completely insulated from them. This led to the development of the "switch," which at first was large and crude and operated manually by humans. Next came the discovery that the electrons inside the element silicon behaved even more peculiarly, sometimes acting as insulators and sometimes as conductors. The term "semiconductor" was coined, and the atomic level switch was developed.

Category Two—THE BOTHERED SCIENTIST

This category includes scientists such as theoretical physicists *cum* philosophers who are equally concerned with *what* happens as well as *how* it happens. Category One scientists, of course, outnumbered those of Category Two. But, all the scientists referred to in this chapter or elsewhere in this book are of the second category. They continued their efforts to understand *how* it happens, and in the process learned a great deal more on *what* happens, which in turn was passed on to the Category One scientists, and the technological revolutions continued unabated.

We are here concerned with the Category Two scientists with regard to the continuation of our story relating to the development of quantum physics.

1926, Max Born — The Probability Wave

The waves in the Schrödinger wave theory were not matter waves, they were waves of probability. The Schrödinger wave was not describing an electron at all, instead it was showing the probability of finding an electron at a particular location, and this probability was equal to the square of the amplitude of the wave.

There are many features of quantum physics that defy common sense and are difficult to believe. There will be many new features still to unfold. The universe, though fourteen billion years old, is considered still in its infancy. There is ample time for discovery. The ultimate aim of science is to reach a perfect understanding of what is reality, mind boggling as it might be when it is fully understood. But, it has to be just a wee bit less mind boggling than the statement: "God created this universe by magic."

The probability wave is one such feature that is extremely difficult to comprehend. Before I venture to talk about the probability wave, it is necessary to talk about a certain experiment called the "Double Slit Experiment."

The Double Slit Experiment

Considered by some scientists as the most beautiful experiment ever…Considered by the scientist Richard Feynman as encapsulating the only mystery of quantum physics: "All of quantum mechanics can be gleaned from carefully thinking through the implications of this single experiment."

The purpose of this experiment is to differentiate between a particle and a wave. It was carried out for the first time in the early nineteenth century by Thomas Young to demonstrate the wave nature of light. The apparatus for this experiment in its simplest form includes:

A source emitting the "item" (such as light, electrons, or stones) to be tested, if this item is a particle or a wave. The source could emit the "item" through a hole in a large screen. Let's call it "screen 1." Another screen, screen 2, with two slits in it through which the "item" in question is supposed to go through. The two slits are parallel to each other and located at some distance apart. A third screen, screen 3, which can be a blank wall acting as a detector where the "item" when it strikes the wall makes a certain pattern. After observing this pattern, the experimenter is able to differentiate if the "item" is a wave or a particle. If it's a wave — we know the characteristics of waves — after passing through the slits, the waves will spread out as a result of diffraction, interfere with each other, and produce a pattern of light and shade. The waves march in step, adding together to make bright patches (in case of light). The waves out of step cancel each other to make dark patches. If the item is made up of particles such as stones and these are hurled through the two slits, you would see the pattern on the detectors as just two piles of stone marks, one behind each hole, as these stones did not interfere with each other as they passed through the two slits. If a slit is wide the stones pass straight through and produce (as more and more stones are passed through) a stripe in the screen the same width as the slit. If the slit is narrow, the stones on the screen register marks sporadically over the entire screen.

Consider now that you are an observer and carrying out an experiment with a beam of electrons for the purpose of determining whether electrons are waves or particles. Screen 3 can in this case be a phosphorescent screen that registers each electron's arrival by producing a flash, and the source could be a hot tungsten filament that boils off electrons. At first, you start the experiment by switching on the source of electrons and then go out of the room and come back after a certain interval of time. When you come back and look at the detector screen, you get your first shock. Here is what you find. Let's call it picture P1:

The electron flashes individually appear as dots on the screen showing them to be particles. However it can also be seen that there are regions, say RD, where the density of dots is enormously high as well, say regions R0, where there are no dots at all. This kind of pattern on the detector screen is similar to that caused by waves; at the same time the regions RD, however dark they may be, clearly show that they are made up of closely spaced dots.

How can this be possible? Could it be that the electrons travelled as waves interfering with each other but arrived as particles on the detector screen and displayed a pattern on the detector screen similar to that attributed to waves?

Next, you carry out the experiment once more. Replace screen 3 with a new phosphorescent screen. Switch on the electron beam, but this time you do not go out of the room. Instead, you watch all the proceedings step by step and see what happens.

Begin with just one electron. It passes through the slit (you do not know which one) and registers a flash on the screen. A dot appears—it's a particle—not a wave. Next, allow about ten electrons to be emitted from the source. They pass through the slits, register flashes on the screen, and ten dots appear. They are particles, not waves, with no clear-cut pattern. This is no big deal, however.

Next, allow about a hundred electrons; you see about a hundred dots on the screen—but look carefully, there is a semblance of a pattern developing. You begin to see regions where there are dots in good numbers, with some regions in between where there are no dots at all. Next, allow about one thousand electrons—one thousand dots—the pattern gets clearer and clearer: regions of large concentrations of dots separated by regions with no dots at all. What in the world is going on?

Next, allow about ten thousand electrons—as many dots appear. The pattern is now crystal clear. It is exactly like P1, which you saw when you carried out the experiment the first time, when you went out for some coffee after switching on the beam of electrons and came back to see the result: a pattern of alternate high density dots (RD) separated by zero dot regions (RO), similar to the pattern of alternate light and shade in Thomas Young's experiment with light.

So it's a wave, is it? But what kind of wave? It's different—it surely is different. But it does look like a wave—even though it's made up of dots. We'll come back to this question soon. Meanwhile, we can formulate the first set of conclusions: The electron is a particle when we observe it in a classical way, but when we look at the non-classical side of the coin we see it as a wave.

What about electromagnetic radiation or light? The picture gets reversed. When we observe it in a classical way, it's a wave, but when we look at the non-classical side of the coin we see it as made up of streams of particles called photons. There are as many as one hundred quintillion photons emitted per second by a sixty-watt bulb, and if we can carry out the Double Slit Experiment with individual photons (with monochromatic light), we observe that the light arrives at the screen in discrete localized units of energy.

This "wave particle duality" is the central idea of quantum mechanics. In the words of Randy Harris (non-classical physics):

"Things may behave as waves or as discrete particles depending on the situation. The "situation" might be an experiment designed to reveal a property of something, or it might be determined simply by the dimensions of a region to which the "thing" is confined. In a classical situation our observation (and any observation is after all an experiment) reveal electromagnetic radiation behaving as waves and massive objects as discrete particles. But when we look at the non-classical side of the coin we see electromagnetic radiation behaving as a collection of discrete particles and massive objects behaving as waves."

The Double Slit Experiment carried out with photons reveals another significant feature of quantum mechanics: the flow of energy is never a continuous process, rather it is in units called "quanta," and a photon is a quanta of light. In the words of Roger Penrose (*The Emperor's New Mind*): "Never is the energy of just half a photon (or any other fraction) received. Light reception is an all or nothing phenomena in photon units. Only whole numbers of photons are ever seen."

But, what exactly happens when the electrons (or photons) travel through the slits and how, or rather when, do electrons (which are apparently particles) decide to interfere with each other to project the wave phenomena? To answer the above question, let's carry out the experiment a few more times, but this time in a slightly different way.

Replace the phosphorescent screen (screen 3) with a new one. On this screen mark out the regions RD and RO precisely at the identical locations corresponding to the earlier experiments. Block one of the two slits. Begin the experiment by switching on the electron source. What happens?

You get your second shock when you see what you see—dots spread out sporadically as if you have done an experiment with stones. And these dots are found in regions RD as well as RO. Of course, the intensity of dots is higher just behind the slit zone, and reduces on either side of the zone, but there is no special preference for region RD over region RO. Now, carry out the experiment for the fourth time in the same way as experiment 3, except that this time you block the other of the two slits. The result is similar to that of experiment 3.

Now, re-collect all four experiments and formulate your own conclusions. In experiments 3 and 4, with one slit blocked in each case, the electrons do not hesitate going through all the regions including RD and RO. But, in experiments 1 and 2, with both slits open, despite an additional choice open to them, the electrons somehow do not travel toward regions RO, but travel in much larger numbers in regions RD. What in the world is going on?

The electrons appear to be fully aware of the entire experimental set up. They _know_ whether the other slit is blocked or not and behave in a different way. In short the electrons seem to _know_ when you are watching them and adjust their behavior accordingly. If the experimental set up was such that screen 2 had only one slit, and that was open, you would have carried out the experiment and observed scattered dots on screen 3—like those created by bullets fired through the slits—and gone home convinced that an electron is a particle—no questions asked.

The Double Slit Experiment—with electrons—was originally just a "thought experiment" (later actually carried out several times) conceived by Richard Feynman sometime in the early 1950s. Indeed, it is a measure of the ingenuity of the human brain that such a "thought" should strike the mind. After all, where was the need to imagine that a electron could also be a wave; there should have been no question that an electron is a true particle. It was already a proven fact. Its charge, mass, and spin were predominantly measurable, and it left tracks in a Wilson cloud chamber. It looks as if the "quantum of genius" or the "advancement of thought" was comparable to that achieved by Leibnitz/Newton, when they discovered calculus independently or Einstein when he discovered the relativity theories. The fact is, however that Richard Feynman, who conceived this thought experiment, did so based on his existing knowledge of quantum physics, and that this thought experiment was just a smart and ingenious way of explaining the various aspects of quantum mechanics in one of a series of lectures (sponsored and broadcast by the BBC), which he gave. Aspects such as Born's statistical interpretation and the Superposition Principle taken together imply an observer created reality, as well as features such as those brought about by "The Uncertainty Principle" (to be explained later) and, of course, the mind-boggling feature revealed by the two words ``P .......... W...``

What are these two words? Let me go back to the experiment and the questions asked: So it's a wave, is it? But what kind of wave? It's different...it surely is different. But it does look like a wave—even though it's made up of dots.

Prepare for the third shock...An astounding one: Ladies and gentlemen—the two words that describe what kind of wave it is are: "p r o b a b i l i t y w a v e." The door to determinism is closed. It was never open. God plays dice after all. Even He cannot tell what an electron is going to do. He only knows the odds and can fix the probabilities.

How was the idea that there is something called a "probability wave" born in the first place? It was born in the mind of a man called Max Born—a masterstroke of imagination...to say the least...one of the all time masterstrokes.

That the electron is a wave was already established. The French scientist Louis de Broglie had earlier theorized by analogy that if light, which was apparently a wave, could act like a particle — photon — as shown by Einstein, then an electron, which was clearly a particle, must sometimes behave like a wave. In a paper he wrote in 1923, he deduced the electron's wavelength. Erwin Schrödinger even devised an equation that defined the shape of the wave — this equation marked the beginning of "wave mechanics." Later experiments demonstrated clearly that electrons exhibited diffraction, thus leaving no doubt whatsoever that true waves were involved. But it was Max Born in the year 1926 who showed that these waves were no ordinary matter waves; they were instead waves of probability. He came out with a stunning revelation that the Broglie-Schrödinger wave function specified the probability of finding the wave at any specific point or location, and that this probability was proportional to the square of the amplitude of the wave. This was a paper on the statistical interpretations of the wave phenomena. Born declared that "The motion of particles follows the laws of probability but probability itself propagates in conformity with the laws of causation."

This aspect is best explained in Roger Penrose in *The Emperor's New Mind*:

"Probabilities do not arise at the minute quantum level of particles, atoms, or molecules — these evolve deterministically — but seemingly, via some mysterious larger-scale action connected with the emergence of a classical world that we can consciously perceive."

To illustrate this, consider a road leading toward a T-junction. If you turn left at the T-junction, you are proceeding toward a fifty thousand–person capacity cricket stadium where a Twenty20 International is about to start and a capacity crowd is expected to see the match. However, if you turn right at the T-junction, you are proceeding toward a ten thousand–person capacity auditorium where a rock concert is due to start at about the same time as the cricket match and a capacity crowd is anticipated there also. For someone who is looking from high above in a helicopter, he sees a huge crowd moving toward the T-junction; he cannot say for sure if a particular individual in the crowd is going to turn left or turn right, but he can very well calculate that the probability of the individual turning left is five out of six, and that of his turning right is one out of six. This is how quantum entities behave. We cannot say what they are going to do but can only fix probabilities.

Indeed Einstein—the epitome of the bothered scientist—was disturbed, and his response to Born's theory was his famous statement "God does not play dice." According to Einstein, "Quantum mechanics calls for a great deal of respect, the theory offers a lot, but an inner voice tells me it is not yet the real thing. I am convinced that God does not throw dice." Born was obviously disappointed with Einstein's comments. But, Born was convinced that it was the probability distribution of events that is determined by quantum theory and not the outcome of specific events.

Born's response to Einstein's remarks was this:

"If God has made the world a perfect mechanism, he has at least conceded so much to our imperfect intellect that in order to predict little parts of it, we need not solve innumerable differential equations but can use dice with fair success."

The probability wave opened the door to indeterminacy. And that door was opened still wider by what happened next.

1927, Werner Heisenberg—The Principle of Uncertainty

"Nothing is more important about the quantum principle than this that it destroys the concept of the world as 'sitting out there,' with the observer safely separated from it by a twenty-centimeter slab of plate glass. Even to observe so miniscule an object as an electron, he must shatter the glass. He must reach in. He must install his chosen measuring equipment. It is up to him to decide whether he shall observe position or momentum. To install the equipment to measure the one prevents and excludes his installing the equipment to measure the other. Moreover the measurement changes the state of the electron. The universe will never afterward be the same. To describe what has happened, one has to cross out the old word "Observer" and put in its place the new word "Participator." In some strange sense the universe is a "Participatory universe."(John Wheeler).

"The mere fact that a phenomenon has a wave nature implies inherent uncertainties in its particle properties. When an electron matter wave passes through a single slit, it spreads out and creates an uncertainty in its momenta" (Randy Harris — *Non Classical Physics*).

"In 1926, the picture of the subatomic world of electrons was one of "Standing Waves" of Schrödinger and "Probability Waves" of Max Born. The hard reality of the electron had already melted to wave-particle duality. In 1927, Werner Heisenberg in an epoch-making discovery, finally showed that not only the electron picture is a blurred one, but that the electron itself is not knowable through any possible scientific experiment" (Swami Jatatminanda — *Modern Physics and Vedanta*).

Without doubt, this leap of imagination regarding the principle of uncertainty by Werner Heisenberg must rank among the top leaps of imagination of all time. How did this happen? Let's get back to Heisenberg's matrix mechanics.

Consider "q" as a matrix, which represents the position of a particle, and "p" as a matrix representing its momentum. To us mortals this mathematics of matrices is difficult to understand, but Heisenberg had no difficulty in realizing that if these two matrices p and q had the property that (p x q) was not equal to (q x p) then it is clear that it is not possible to simultaneously measure both the properties with complete accuracy. Heisenberg used a Gedanken experiment (a thought experiment) to explain this phenomenon. He imagined that if a super-powerful gamma ray from a super microscope is focused on the fast-moving electron, no doubt the electron is illuminated, but it is violently knocked out of its orbit, and changes its direction and speed (which defines the momentum). This happens in an uncontrollable and unpredictable way, and though an apparatus may well make a simultaneous measurement of position and momentum, there will be inherent uncertainties in the said measurement. Heisenberg even derived an equation:

$$Qx*Qy > \text{or} = h$$

Where Qx is the uncertainty in the position measurement and Qy is the uncertainty in the momentum measurement. What is "h?" Believe it or not, it is the same old Plank's constant (6.63 x 10^{-27} erg/sec.).

What exactly do we mean by the term "uncertainty?" and how can we say with a measure of certainty that a certain measurement is uncertain, let alone arrive at the mind-boggling equation derived above?

Consider a single measurement made on determining "q" the position and "p" the momentum of a particle by pressing a button of the apparatus that does the measurement. Who in the world can say that this measurement is wrong or inaccurate, and if it is inaccurate, what is the extent of uncertainty in our knowledge of its accuracy? The fact is, with a single measurement, we cannot evaluate uncertainty as there is nothing to compare it with. Only if we carry out large numbers of experiments can we ascertain the quantum of uncertainty by statistical methods. For this we need to define what uncertainty is. The most appropriate definition of uncertainty was arrived at by utilizing the "standard deviation" criteria of statistics and taking this uncertainty as equal to the square root of the mean of the squares of the deviations of the value from the mean value, divided by the mean value.

To illustrate this with an example, let me take two sets of measurements of, for example, the temperature (in degrees centigrade) inside a room during a one-hour period at two different times and compare the uncertainties of each set. Consider the measurement values as under:

First set (Taken between 3 p.m. and 4 p.m.): 15.4, 15.3, 12.9, 13.6, 12.8

Second set (Taken between 9 p.m. and 10 p.m.):11.0, 11.8, 11.2, 10.6, 10.4.

In the first set, the mean value is 14 degrees and the uncertainty is worked out as 0.081. In the second set, the mean value is 11 degrees and the uncertainty is 0.045, thus showing that the uncertainty of the first set is a little less than twice that of the second set.

Heisenberg's equation is applicable only if a large number of measurements are made and a statistical average determined. The quantum of uncertainty in a measurement starts increasing stupendously as the size of the object gets smaller and smaller and moves from the classical to the quantum domain. As the value of Planck's constant "h" is not zero, if one of the parameters, position or momentum, is definitely known, then the uncertainty pertaining to the other parameter becomes infinite. A close look at this aspect will show that empty space is not empty at all and there is a sea of activity going on all the time.

Impact on Science:

Heisenberg's uncertainty principle shook the very foundations of science. In fact, two such pillars constituting the main foundation of science were demolished.

Newton's laws of motion were found not applicable at the quantum level. Physics was henceforth divided into two parts: classical physics for macroscopic objects of size "z" and above, where Newton's laws are applicable and quantum physics for microscopic objects of size below "z" where Newton's laws are not applicable, with no clear cut dividing line that could define z, except by quantum of required accuracy.

We cannot describe reality in an objective way, in the sense that we cannot observe anything without changing it. The scientist, who does the observation and takes the measurement, cannot be treated as a detached /indifferent observer; he is in fact very much a participator or you might say an actor in the drama of this quantum play.

There Is Nothing in the World Called Nothing:

The uncertainty principle applies even in an empty region of space such as a vacuum; it tells us that energy and momentum are uncertain.

"It's as if this region of empty space is a compulsive borrower of energy and momentum, constantly extracting 'loans' from the universe and subsequently 'paying them back'" (Brian Greene — *The Elegant Universe*.

If this uncertainty principle is combined with Einstein's ultimate convertible currency equation $e = mc^2$ it reveals that energy can turn into matter and vice versa, and if this quantum of fluctuation of the energy is big enough, it will create an electron and its antimatter called a positron from nothing at all, and the two of them will annihilate themselves in no time at all. Hence this "Creation and Annihilation — Creation and Annihilation — Creation and Annihilation" goes on all the time in empty space, which — notwithstanding all this violent activity — remains calm and placid for the simple reason that the amount of loan borrowed is precisely equal to the amount repayed.

The philosophers who will be invited here to discuss the philosophical implications of quantum theory to try and answer the "How it happens" question are many in number and indeed they have a lot to say, and so a full chapter — chapter six — will be devoted exclusively for this purpose. However, in this section I am inviting one of them (although there will be one more, Roger Penrose, near the end of this chapter) — James Jeans — to sum up what according to him is the picture that emerges from what we have learned regarding the features of quantum theory, in particular the principle of uncertainty/indeterminacy:

"...The fundamental laws of nature do not control the phenomena directly. We must picture them as operating in a substratum of which we can form no mental picture...Events in this substratum are accompanied by events in the world of phenomena, which we represent in space and time, but the substratum and the phenomenal world together do not form a complete world in itself, which we can observe objectively without disturbing it. The complete closed world consists of three parts — substratum, phenomenal world, and observer" (Amaury de Riencourt — *The Eye of the Shiva*).

We now move on to another feature of quantum physics and the principle subject of this chapter, quantum entanglement.

Quantum Entanglement

As far as the question of fastening the safety belt is concerned, let me tell you that this is the turbulent portion of your flight. If you do not understand this section, read it again. And again…and again…and again. If you still do not understand (comprehend is the right word) it, you are in good company. Let me tell you something: No one understands it *yet*.

If this feature of quantum physics is understood by scientists, we can very well say that reality is understood and that's the end of science. When I say the word "understand," I mean it in the sense "understand how it happens." Understanding in the sense "what happens" is no big deal; this is understood by many. In fact, some category-1 scientists of the world, such as some working at Microsoft, Intel, and other IT companies are already doing intensive research on the subject, in their quest to be the first ones to usher in quantum computation. (I hope they do not succeed until about three hundred years from now, considered the danger zone in regard to the probable "Collapse of Civilization.")

Meanwhile, all you can do is to "get used to it" and improve your understanding from a level "a" to a certain level "b," where you can start feeling the turbulence, and then read it again and again along with the chapter on the "six words," and then again read all the previous chapters, and repeat the process till you reach a certain level "c" of understanding where you can formulate your own philosophy, in a way that this philosophy can act as a safety belt to take care of this turbulence including all the turbulence in your life.

Part of the conclusion that will be drawn near the end of this chapter will be something like this:

"By a remarkable act of sleight of hand, the God that plays dice has made sure that while I or anyone cannot pass any information at a speed faster than light, the component parts of my body made up of quantum entities can and do interact instantaneously with their counterparts, with whom they had once interacted, located billions of miles away, perhaps in some other galaxy."

Einstein-Bohr Debate and the EPR Paradox

I mentioned earlier that two categories of scientists emerged as a result of the understanding of quantum theory, being the "indifferent scientist" and the "bothered scientist." Now, after Heisenberg's principle of uncertainty or rather the principle of indeterminacy, these bothered scientists concerned with the interpretations of quantum mechanics were themselves divided into two categories.

The first of these two categories—let's call it Category 2A: somewhat bothered and somewhat indifferent—was the larger of the two camps led by Niels Bohr. According to Bohr, "particle" and "wave" were complimentary concepts giving different representations of the same object. His complimentarity principle described quantum entities like electrons in ways that are mutually exclusive, i.e., either as waves or particles depending on how the experimental arrangement was set up or which aspect was required to be measured. At Copenhagen, the scientists led by Bohr, including Heisenberg, Max Born, and many others, took stock of the situation and arrived at the famous "Copenhagen interpretation of quantum mechanics," taking into account Heisenberg's uncertainty principle, Bohr's complimentarity principle, Max Born's statistical interpretation—probability wave—as well as the phenomena pertaining to the "observer created reality," such as the identification of the state vector with "knowledge of the system," implying that the observer who does the experiment just seems to acquire a certain knowledge of the system. In other words, you might say that the so called "collapse of the wave function" would imply that it is the "knowledge" that does the jumping and not the physics of the system.

The essence of the Copenhagen Interpretation (CI) was providing a connection between the mathematics of the formalism of quantum physics, and the physical world, or rather testing the formalism by verifying its predictions with experimental results. All this is with regard to the observed phenomena. With regard to something, which is not observed, the CI takes the stand that it is meaningless to say if it exists or it does not exist. In short CI rejects determinism, accepting the statistical nature of reality. It also rejects objectivity, accepting material reality as depending on how we choose to observe it.

Category 2B scientists, the "much bothered" category, were fewer in number. Led by Einstein they included Schrödinger, de Broglie, Max Planck, etc. Einstein was not in agreement with the viewpoint that the world must be actually observed to be objective, nor was he happy with the indeterminism of quantum theory. A debate ensued between Einstein and Bohr that lasted for several years, considered by many as the greatest ever debate in intellectual history, between two of the greatest minds in the world, on such deep issues as the understanding of reality and this strange world of ours, in short on physics itself.

In my opinion, the debate between Einstein and Bohr, running into several rounds and culminating in the EPR paradox, is the ultimate in the "uniqueness of the human brain," which in turn is the uniqueness of all uniquenesses — The World Cup of all World Cups.

That the "going" is tough is an understatement and here, on my part, I will dwell on it only to a limited extent, covering broadly the sequence of events leading to the EPR paradox. For comprehensive coverage, and to those readers whose capacity to understand the subject is reasonably high, I recommend that they should read Andrew Whitaker's book *Einstein, Bohr, and the Quantum Dilemma,* which in my opinion "takes the cake" among all books written on the subject.

One after the other, Einstein came out with stunning ideas to counter the uncertainty principle, and each time Bohr came out (albeit after some sleepless nights) with equally stunning and convincing replies to prove Einstein wrong. With each episode, the mutual respect, admiration, and love for each other kept growing. Einstein never hesitated to accept his mistake each time it was pointed out to him. With every episode, his admiration for the quantum theory kept increasing, yet his conviction that it was an incomplete theory remained intact until the end. One such dramatic encounter between the two giants is briefly described below. Very few writers on quantum physics miss out on this:

1930 — The Sixth Solvay Conference — The Clock in the Box Experiment

Some of the most significant discussions between the two greats did not take place in the conference rooms. Like this one, they happened rather in hotel dining rooms. Einstein started off with an ingenious thought experiment:

Imagine a box with a clock inside the box. The box is closed with just a minute hole in one of its walls or maybe at the top. This hole can be opened or closed by a shutter, which is controlled by the clock inside. The box is completely filled with radiation. The box is weighed and then the shutter is set to open for the briefest of intervals during which just a single photon is allowed to escape. The box is then weighed again. The difference gives the photon's weight from which its energy can be determined. The time corresponding to the opening of the shutter is the time of passage. Hence both time and energy are determined to an arbitrary accuracy, which is in conflict with the energy-time uncertainty principle.

Bohr was shocked. It could not be true; if Einstein was right this could be the end of physics, but he had no answer at the time. Scientists who were present in the room could never forget the expression on Bohr's face. Professor Rosenfeld, who was present at the scene, recalled: "I shall never forget the vision of the two antagonists leaving the club, Einstein a tall majestic figure, walking quietly with a somewhat ironical smile and Bohr trotting near him, very excited."

Bohr had a sleepless night, but before dawn he found the answer, and met Einstein the same morning. Astonishingly, it was Einstein's general relativity itself that was utilized by Bohr to prove his point. The reduction in the mass of the box caused the surrounding gravitational field to slow down, causing a change in the scale of time in a way that the product of the two uncertainties (time and energy) was fully in accordance with the uncertainty principle, i.e., equal to the minimum value predicted. However, when the time measurement was precise, the energy was completely uncertain. Thus not only was it that the uncertainty principle could not be proved wrong, but also in the process the uncertainty equation was verified.

As I mentioned earlier, some readers may find the going "heavy" and "yielding" — don't worry, I am not much better off. I'll try once more, or rather let Andrew Whitaker do the talking:

"Assume the box is weighed using a spring balance, and the result is obtained from the position on the scale of the pointer attached to the box. Now we must accept that this 'position' must have an uncertainty 'dy,' which is related to that in the momentum of the box 'dp', by the uncertainty principle. To obtain some information about dp, Bohr imagined the limit of accuracy in the measurement of the mass being 'dm.' Then dp could not be larger than the momentum given to dm in period T, which is just the product of dm, T, and the gravitational constant g.

Now came Bohr's masterstroke. According to general relativity, a clock that is moved in the direction of a gravitational field will change its rate. Thus the uncertainty dy in the position of the pointer, and hence of the box, gives rise to an uncertainty dt in the time interval T. The uncertainty in the momentum dp, corresponds to an uncertainty in the energy de, and when the detail sums are done, one obtains the result that the product of dt and de is at least as great as Plank's constant h — just the time-energy uncertainty principle." (*Einstein, Bohr, and the Quantum Dilemma*).

Einstein was convinced. It is a measure of Einstein's character that he was always gracious in accepting his mistake. In this particular case he realized that he was wrong (that he had not taken into account the peculiar effect of gravity) even before Bohr could complete his explanation and he rather helped Bohr in elucidating his response. But the extent to which Einstein was convinced was limited to the realization that the arguments put forth by him to disprove the uncertainty principle were flawed, and he needed a different approach and a more subtle and convincing set of arguments to drive home the point that quantum theory, despite its profound beauty, its general validity, and its huge success with all its practical applications, was somehow incomplete, and that there must be certain unknown properties of the system, or, in other words, certain "hidden variables," which were not obvious at the time, but are bound to be discovered in the future that should explain this discrepancy.

Who won the debate between Einstein and Bohr? Nine out of ten scientists (even of the bothered variety — the indifferent ones concerned themselves only with the practical applications and did not feel the need to think beyond the standard Copenhagen interpretation) considered Bohr as the outright winner, at least up to the time Einstein presented the EPR paradox. There were many, however, who did not subscribe to this viewpoint. The trouble is that we are still far, far away from the truth on "what is reality" and "what is quantum physics." This picture of reality and this confusion regarding who was winning the debate was further complicated in a *big* way by the EPR paradox leading to this "quantum entanglement."

1935, Einstein-Podolsky-Rosen: EPR Paradox; Quantum Entanglement...The Debate Continues

Before I begin talking about the EPR experiment, let me again tell the reader that the "going" is still heavy (with regard to the understanding of these phenomena), and it will remain "heavy" until this chapter is concluded. And this "going" is heavy not just for you and me but even for all the great scientists. In fact, even for the two great scientists Einstein and Bohr, the "going" is far from good or firm; at best you might call it "soft." For if one of them is on firm ground, then for the other the ground or the going is "yielding." For after all, it was a debate, and it's still not resolved who was on firm ground. When the EPR paradox was presented, the scales seemed to have tilted in favor of Einstein. Thirty years later when Bell presented his "inequality concept," and Alain Aspect performed a certain experiment, the scales tilted again in favor of Bohr (or did they?). Well, you can judge it for yourself when the discussion is over and before the philosophical overview is presented. My own view is that:

"Neither Einstein nor Bohr lost the debate."

The long and short of it is that if neither of these two gentlemen were on completely firm ground, then how can we mortals hope to find the going "good and firm." Keeping this aspect in mind, just read on and get used to the "going," and think of this as a story — the story of quantum physics — the story that led to the "six words" — the six words that led to this autobiography — the six words that could be the "trump card" that might save the collapse of this civilization from "self-destruction."

What exactly was the EPR paradox? It is said that the word "upset" was coined in the dictionary when Man o' War, considered as the greatest thoroughbred race horse of all time, was beaten by a horse called "Upset." In the same way, if some people asked me to make one guess and tell them how the word "brilliant" might have been coined, I would have to say that the front runner would be the EPR paper.

Einstein, Podolsky, and Rosen were the three scientists who in 1935 wrote a paper known as the EPR Paper, describing a thought experiment called the EPR experiment, and called by many as the EPR paradox. This paper could arguably be the most discussed paper of all time. In fact, fifty years later in the year 1985, there was a world conference held in Singapore to mark the fiftieth anniversary of the EPR paper. At this conference there was even a debate as to whether the word "paradox" was appropriate. The word, in fact, was used by Einstein himself in the context that if quantum theory is true and complete, then it leads to a certain paradox, a certain something that is impossible to imagine, a certain something that defies determinism, a certain nonlocal action defying the laws of causation, in short a certain "spooky action at a distance" — and all these things are absurdly impossible. It means only one thing that quantum theory is as yet an incomplete theory.

One of the features of quantum theory, including the uncertainty principle, is that "not only is it impossible to simultaneously measure the position and momentum of a particle with a degree of certainty, there is no objective reality to the existence of these features — they do not exist at all till such time as a measurement is made on them."

The EPR thought experiment intends to disprove the above statement. The argument put forth by Einstein on which the paper was based was this:

"If without in any way disturbing a system, we can predict with certainty (i.e., with a probability equal to unity), the value of a physical quantity, then there exists an element of physical reality corresponding to that physical quantity" ("Some strangeness in the proportion" — proceedings of a seminar, edited by Harry Woolf).

Consider a particle "p", which is at rest, decaying into two particles p1 and p2, and flying off in opposite directions, which means no more at rest. Prior to getting separated, the wave function of the combined system can be taken as "tangled," which means the two particles were in "interaction" with each other, and hence their future behavior is correlated. From the conservation of momentum, the total momentum, which was zero in the beginning, remains unaltered at zero. Hence, if we measure the momentum m1 of the particle p1, we know immediately the momentum m2 of the particle p2, as the total momentum is conserved. From the definition of momentum, which is nothing but the product of mass x velocity, where velocity is nothing but the speed in a specific direction, it becomes clear that the positions of the two particles are related, and so if we measure the position q1 of particle p1 from the origin, we immediately know the position q2 of particle p2 from the same origin. Now, when we measure the momentum of particle p1 with a degree of certainty, we know that, as per the uncertainty principle, the act of measurement has resulted in a disturbance, which has caused an uncertainty in its position. But, there was no such disturbance caused to the second particle as there was no direct act of measurement, both the momentum and position of particle p2 having been determined by deduction, using the criterion that the combined momentum and the combined (relative) position are known and that knowledge pertaining to the combined status is not denied by the uncertainty principle. Hence, we have been able to measure precisely, both the momentum and the position of particle p2, which is against the uncertainty principle. But, this conclusion has been reached after taking local causality into consideration, which means local causality was not violated, by taking into account that no such information was passed on to particle 2 regarding the act of measurement. That particle p2 had no way of knowing whether the precision measurement was done on the momentum or the position. In short, the only way the quantum theory can remain unblemished is on the assumption that local causality is violated, which means that there was nonlocal action or action without causation. As this is absurd and impossible to believe, Einstein opined that quantum theory is incomplete.

It is, of course, to be assumed that the two particles were widely separated, perhaps a few light seconds apart, to ensure that the information on the act of measurement of momentum on particle p1 could not instantaneously be communicated to particle p2. Accordingly, it can be inferred that without causing any disturbance to p2, its momentum m2 was determined precisely, and if this particle p2 had a precise value of its momentum at the time when a measurement was performed on p1, it must have always had a precise value of momentum even before that measurement, since at no stage was any disturbance caused to particle p2. Accordingly, quantum theory, which denies objectivity, remains an incomplete theory.

The EPR paper shook the world of physics and caused an enormous stir among physicists and philosophers. The effect on Bohr was like a bolt from the blue. Of course, he still maintained that the position and momentum of the particle p2 have no objective meaning until they are directly observed, in which case they will obey the uncertainty principle in agreement with quantum theory. But it was clear to everyone that the Copenhagen interpretation had lost a great deal of its sheen. It received another blow when Schrödinger came out with his thought experiment—the famous "Schrödinger's cat"—which clearly and cleverly demonstrated that the quantum weirdness at the level of atoms could creep up to the macroscopic level of a cat. To visualize this, consider a cat in a closed chamber having an experimental set up that can be manipulated to cause an atom to decay with a 50 percent probability in a fixed interval of time, and that if this atom does decay, it converts the air inside to a poisonous gas that kills the cat, which means that the state of superposition of the atom having decayed or not decayed gets converted to the state of superposition of a cat, which is now to be considered both dead and alive at the same time until such time as an intelligent observer opens the chamber and collapses the wave function of not just an atom but an entire cat made of trillions of atoms. And this is not all. In the eyes of the people sitting in a meeting in a conference hall, until the observer communicates to the members in the meeting the result of his observation, he himself must be considered in a state of superposition. Notwithstanding these onslaughts, the supporters of the Copenhagen interpretation remained unperturbed. Their stand: You are talking of cats; the moon itself does not exist, if no one is looking at it.

And so the debate continued…and it still continues. Is quantum theory complete or is it incomplete? For thirty years the question remained unresolved. And then came another scientist called Bell…and did he indeed ring the bell with his staggering leap of imagination — the astounding Inequality.

1965, John C. Bell — Bell's Inequality; 1982, Alain Aspect — Aspect Experiment (4-004)

BRIEF OUTLINE OF BELL's IDEAS:

Two sets of measurements need to be carried out on a pair of quantum-level particles that were in contact at some time, but are now separated. The entangled quantum state is written down for the four, spin one-half particles, and the expectation value of the products of certain binary measurements performed on the individual particles is calculated. It is then worked out on the basis of the probability theory that if local causality is considered while determining the expectation value, it produces a contradiction.

This contradiction is called Bell's Inequality. If this contradiction is satisfied, it confirms locality. If this contradiction or inequality is violated, it confirms nonlocal action.

In *Cosmic Code* by Heinz Pagel, two experiments have been described. One is for classical level objects—special nail guns shooting nails sideways with their long axis perpendicular to the direction of motion, where Bell's inequality is satisfied confirming local causality. The other is for quantum-level particles—positronium atoms decaying into photons and then travelling in opposite directions, where Bell's inequality is violated confirming nonlocal action.

Recall the great adventures of Sherlock Holmes, or the books of Agatha Christie, Edgar Wallace, or Dan Brown, or any other book, which caused you to say "Wow, that was clever" and then judge for yourself if that quantum of cleverness could match the quantum attributed to Bell and his inequality. And can we not attribute this adjective "clever" in equal measure to Roger Penrose and Heinz Pagels in capturing these ideas and putting them across in such an ingenious manner. Also, in equal measure, do we recognize Andrew Whittaker in first understanding and then describing a complete historical review of the past sequence of ideas from de Broglie's "pilot wave" concept to Bohm's "hidden variable" concept and many others leading up to Bell's ideas.

Many experiments were actually carried out by the scientists in the eighties. In 1982, an experiment carried out by the French scientist Alain Aspect on photons up to a separation distance of about fifteen kilometers successfully demonstrated that Bell's Inequality gets violated at quantum levels, thus confirming "nonlocal action" (NLA) or "quantum entanglement" (QE).

Quantum theory stood vindicated. However, the question of whether quantum theory is complete or incomplete still remains unanswered. But the EPR paper, which was designed to prove the incompleteness of quantum theory as it led to a paradox called "nonlocal action," could not prove it. Instead this NLA or QE is here to stay and emerges as another feature (mind boggling, no doubt) of quantum mechanics. Indeed we have to live with this "action at a distance," whether spooky or not spooky.

But can we make use of this QE to send messages at speeds faster than light? Coming back to the adjective "clever," let me tell you that none is cleverer than the God who plays dice. He has played a marvelous trick on us by avoiding real time, nonlocal influences. If you read the referred cosmic code chapter carefully, and also a number of times, then at some stage and at some point of time, a certain flash of understanding may strike you, and you will realize that the answer to the above question is: No. We cannot.

The argument that is used for this purpose is that: A random sequence altered randomly remains a random sequence. At this stage I will not go into the details except to conclude that: God is a Mathematics. Indeed "mathematics" is the President and the name of the company is "The universe."

Mathematics designs the universe by creating laws and tells physics to execute them. Laws such as the "uncertainty principle" where "mathematics"—the President—plays such a trick that it does not allow the quanta to get created out of nothing except for the shortest possible—and thus irrelevant—period of time. Laws such as "quantum entanglement" where "mathematics"—the President—plays such a trick that it permits entanglement but does not permit information to be sent faster than at speed of light. And, both these tricks have been played out by the mathematician, by incorporating a certain randomness in the nature of reality. In the former case the vacuum randomly fluctuates between being and nothingness, and in the latter case the mathematician keeps shuffling the deck of nature in such a way that the randomness remains…intact.

I should go as far as to say that "statistics" is the Vice President.

And then at the center of everything, there is this thing called "equivalence," the cleverest of all the laws. It goes without saying that this law is out and out mathematics and physics is just dancing to its tune.

Who won the debate, was it Einstein or was it Bohr? This question too remains unanswered still. At present all we can say is: Neither Einstein nor Bohr lost the debate.

To the question whether QE has any practical applications, "mindboggling" is the word that comes to the mind. Scientists of the category called "indifferent" are already working on it and quantum computation is looming large on the horizon…it could also hasten the collapse of civilization.

2003, Six Words…This Autobiography (4 -005)

The turbulence of the mind continued…

What's going on in this world? How do we explain all these things?

Non-objectivity…The observer created reality…Particle wave duality…The collapse of the wave function…Probability wave… Electron knowing when it is observed and behaving differently, even knowing when it is going to be observed in the future… Quantum entities at two places at the same time…most of all this nonlocal action…this "quantum entanglement…"

The turbulence of the mind continued…

And then one night, all of a sudden the collision was felt. The six words had struck. The turbulence seemed to subside somewhat…

The great scientist John Wheeler once said: "In time to come, a single simple sentence will explain the strangeness of the universe and as we say that sentence to each other we'll say 'Oh how beautiful. How could we have missed it all that time?'"

Could this sentence of six words be that single simple sentence that John Wheeler thought?

It was thinking about all these weird peculiarities of quantum physics that seemed to have triggered the collision. If electrons can be at two different places simultaneously, as shown in the double slit experiment, then why not mind? If electrons could entangle with each other with no consideration for relativity and the laws of causation, then why not mind?

These striking thoughts were the culmination of more than fifty years of "thinking…asking questions…thinking…trying to find answers…thinking…remaining in turbulence."

I tried to go to the past and recollect some of these thoughts.

It all began at the age of ten. The year was 1951. I have a distinct memory of the precise moment when it happened. I do not recollect the interactions that caused it. Something entered the mind and caused some turbulence, a sudden realization of something unusual, which was not taught by the teachers nor by the elders. Suddenly I became a thinker and remained in that category ever since. I started asking the "Why" type of questions... and answering them myself...questions like why A can jump five feet and B can jump only four feet. Why A is rich, and B is poor. Why A was considered good and B was considered bad. I began to understand the law of causation...

And then came 1957. I was in college. I learned the law of probability...and felt comfortable. I read books on astronomy and learned about the vastness of the universe...and felt even more comfortable. I read Thomas Hardy's *Return of the Native* and a certain sentence relating to the rocking of a chair replenished further my understanding of the laws of causation. I continued to think and have remained in nice turbulence ever since...

And then came 1988, my first interaction with quantum physics or rather the philosophical implications of quantum physics through a book entitled *Modern Physics and Vedanta* by Swami Jitatmananda: incredible magnitude of interest created in the subject...accumulation of books on the subject...tremendous struggle in trying to understand the subject...countless hours of getting used to the subject and its complexities...incredible scratching of the head...intense exhilaration each time some semblance of understanding materialized...and so on...

And now these last few months and this phenomenon called "quantum entanglement"...this "spooky action at a distance"... and all these tricks played on us by the God who plays dice... tricks that however are necessary and are in conformity with the anthropic principle...back to the "six words." The turbulence, which had subsided, started all over again. Was I privy to something no one else knew in the world? It could not be possible. How will I cope with it? What exactly should I do?

And then the very next day, by a remarkable act of sheer coincidence, the very next morning, I read an extract from the book *What is Life?* by Erwin Schrödinger, in a book titled *Quantum Questions*, edited By Ken Wilber and gifted to me some time back by my son. The extract contained a certain set of ten words that had precisely the same meaning as the "six words" in question. And then I remembered having read something about an equation attributed to the Upanishads, in *Modern Physics and Vedanta*, which was intricately related to the "ten/six" words. For a brief moment I was a bit downcast with the realization that I was no longer the only person privy to this idea, but later felt even better with the realization that a scientist of the caliber of Schrödinger was of the same view, and the gifted minds of the Vedic period thought on the same lines albeit by intuition, all of which has added a degree of authenticity to these "six words."

A certain ten seconds while reading about the double slit experiment reached the top of my list of the top ten "ten seconds" of my life, later to be relegated to the third spot. The two events described above. The one corresponding to the "night before" and the other corresponding to the "morning after" are the ones that occupy the first two places.

PHILOSOPHICAL OVERVIEW

"Probabilities do not arise at the minute quantum level of particles, atoms or molecules — these evolve deterministically — but seemingly via some mysterious larger scale action connected with the emergence of a classical world that we can consciously perceive." (Roger Penrose — *The Emperor's New Mind*)

The expectation of a classical-level God made up of atoms and molecules and having mass, leads us to a state of mind where we have no choice but to ask the question, are we in good hands? Can such a classical-level God be considered omnipotent or omnipresence? How can this be possible? He would have to occupy a certain limited region of time and space, however large it may be, and be subject to interactions and forces caused by the world outside this region, on which he can have no control whatsoever unless he breaks the laws of physics created by him. If he sits in judgment in a limited region of space in, say, the Andromeda

Galaxy and keeps a comprehensive record of the deeds of all beings on the planet Earth or any other planet in any other part of the universe and then rewards and punishes accordingly, he will have to violate Einstein's special theory of relativity and Newton's laws of motion (at the classical level) and many other laws of physics all created by Him, in a continuous manner. The trouble is we don't see this happening. To think that such a classical-level God will come to our rescue and prevent civilization from self-destruction will be wishful thinking in the extreme. That civilization is distinctly on the road toward collapse is not in doubt and so also is the fact that at present nothing is happening to prevent this. (4-003)

On the other hand, the expectation of a quantum-level God — who is not required to follow the laws of classical physics — and who is conscious and omnipresent in the form of a cosmic consciousness, having created the laws of physics, including the laws of probability and most importantly the anthropic principle is easier to believe. But, in that case, Roger Penrose's statement referred above must be brought to mind. The conclusion is inescapable: There is no other way to prevent such a collapse of civilization from self-destruction except to create such probabilities at the classical level so as to enable the quantum-level God to come to our rescue.

In the eyes of the quantum-level God as an observer, he sees the scientist and the theologian as made up of the same quantum entities, atoms, and molecules, who could easily have changed places had their interactions been different. In principle, he sees no conflict at all between the science and religion. Their roles in understanding reality are complimentary. The former tries to answer the question "How we are here?" by experimental observations and applying the rules of science (physics and mathematics). The latter tries to answer the question "Why we are here?" by intuitive thinking, by taking into consideration the updated status of science, making sure that all the scientific truths acquired and established until that time are not rejected, while at the same time firmly believing in the existence of God until such time as (and even after that) science is capable of supplying answers to the ultimate questions about why things exist and what is their purpose.

God exists at the quantum level, like an "intelligent field" pervading all space, and creates the laws of science in a way that He cannot violate these laws himself. These laws include the law of probability, which too He cannot violate, for He too can just fix probabilities and cannot predict the outcome of an event with certainty. Yet He does all this with a sleight of His hand in a way that leads toward a consciousness that can understand His existence as well as the existence of the universe. All this He manages to do (the sleight of hand part) by creating a certain principle that provides certain mathematics constants to all the laws of science, resulting in full obedience of all the laws including the law of probability, and it is this principle, which we call the anthropic principle.

This then is our religion and everybody's religion. There are no Hindus, no Muslims, no Christians. There are no Indians, no Pakistanis, no Chinese, no Americans. There are no Democrats, no Republicans. We are all made up of the same quantum particles and become what we are as a result of the interactions of the world. That this is a good life and deep thinking and analysis will reveal the "six words," which in turn will reveal how the God that plays dice keeps shuffling the cards in a way that justice is not denied to anyone in the world. There is no need, or justification, or requirement, to take revenge or cause hurt, and no need for nations to remain in conflict with other nations. But this civilization is indeed collapsing and moving steadfastly on the road toward self-destruction, and as yet nothing is happening to reverse the trend. This is a disturbing aspect, and this autobiography is written to create an awareness about it and to drive home the point that it is up to us "to create the appropriate probabilities" to save ourselves from extinction and that there is no such "classical-level" God who will come to our rescue.

**

ENDNOTES

4-001:

If an experiment is carried out on certain electrons or certain other quantum entities in a laboratory in Bangalore, then there exist certain other electrons or quantum entities — which were once entangled with the former — and which may now be in California, or perhaps in the Andromeda Galaxy, which become instantly aware of the experiment, and their behavior is comprehensively related to that experiment carried out perhaps billions of miles away.

4-002:

For further reading:

1) *The Cosmic Code* by Heinz Pagels

2) *The Emperor's New Mind* by Roger Penrose

3) *The Road to Reality* by Roger Penrose

4) *Quantum A to Z* by John Gribben

5) *In Search of Schrödinger's Cat* by John Gribben

6) *Einstein, Bohr, and the Quantum Dilemma* by Andrew Whitaker

7) *Nonclassical Physics* by Randy Harris

8) *Modern Physics and Vedanta* by Swami Jitatmananda

9) *The Eye of the Shiva* by Amaury de Reincourt

10) *Some Strangeness in the Proportion* edited by Harry Woolf

11) *The Tao of Physics by* Fritjof Capra

12) *The Elegant Universe* by Brian Greene

4-003:

Refer to chapters three and six for a detailed discussion on this.

4-004:

To understand Bell's ideas in full or even to an extent that one can write about it in one's own words is an extraordinarily difficult task for me — an engineer who, as you all know, is considered by the world as belonging to the category "an indifferent scientist" — with my limited capabilities, or you might say with the available quality and configuration of the brain cells supplied to me by the God that plays dice. In a subsequent edition I might be in a position to do so, provided I commit myself to a strenuous schedule comprising of:

Interacting with an expert brain analyst and finding an answer to the question: "Can understanding of the brain's functioning improve understanding of how the brain's functioning can be improved?" If the answer is affirmative, then it is a matter of determining the required course of action and then executing that action within a time frame of, say, six months to equip myself with the most appropriate configuration.

Studying all the available literature on the subject and developing an in-depth understanding of it and then composing an appropriate thought experiment, which would be 1) fairly easy to understand by the reader and 2) at the same time having at least a portion of the spellbinding effect that I experienced when I read the following:

Roger Penrose's *Road To Reality*, chapter on "entangled quantum states"

Heinz Pagel's *Cosmic Code*, chapter on "Bell's inequality" (This, of course, is my favorite.)

Andrew Whitaker's *Einstein, Bohr, and the Quantum Dilemma*, chapter on "Bohm, Bell, and experimental philosophy"

These are "must read" chapters, even at the cost of an incredible quantum of scratching of the head.

4-005:

The six words can be determined from the twenty statements given at the beginning of chapter SixE (hint given).

Chapter Five

THE INTERACTIONS OF THE WORLD

**"IT WAS NOT IN MY HANDS...TO BE DETERMINED...
TO BE A DETERMINIST..."**

*THERE IS NO DOUBT THAT THIS IS A DESIGNED
UNIVERSE...DESIGNED WITH GREAT INGENUITY BY THE
SUPERCONSCIOUSNESS OF THE PREVIOUS UNIVERSE...
BUT THE LAW OF PROBABILITY HAS MADE THE UNIVERSE
A HUGE JOKE...AND THAT IN FACT IS THE BEST PART OF IT.*

How big is the joke? This can be gauged from the realization that
it is possible — at least theoretically — that a sequence of events
beginning with a simple mosquito bite can alter the history of
our planet...and if we believe that our technological civilization
will some time in the future be advanced enough to colonize the
galaxy...then it can be said that the same mosquito bite alters the
history of the Milky Way. (5-001)

All events major or minor go back ultimately to a trivial cause.
It is possible that a plain and simple walk in the garden can be
the starting point in a chain of cause-and-effect sequence that can

lead to the extinction of the human race—or the "prevention of such extinction."

A certain equation developed by Einstein might have created probabilities for the civilization to collapse and later a certain letter written by Einstein to President Roosevelt might have created probabilities that prevented such collapse. The said equation is still looming large and probabilities need to be created to ensure that the equation is used for the development of the world and not for its destruction. Each and every event or phenomena—at the classical level—including the writing of this chapter or your reading of it, is the consequence of only *one* sequence of infinity of cause-and-effect events going back to the Big Bang itself. However, each of these events has the potential to initiate and diversify into several altogether new sequences of events.

This then is the law of causation.

As an example I choose a "sequence" connecting Einstein's relativity with the "confidence vote in the Indian Parliament" that took place on July 22, 2008.

Einstein Nukes The Indian Political System:

"In a parallel universe sometime in July 2008, some scientists and philosophers, in different parts of the world, tried to analyze what was going on in connection with a certain confidence vote in the Indian Parliament, making use of these happenings to try to explain some of the features of physics, both quantum and classical by highlighting the probabilistic aspect of the former and the strictly deterministic aspect of the latter.

At Mumbai University a certain professor who is a follower of the standard Copenhagen interpretation of quantum physics (5-002), in a lecture he gave on July 21 (one day before the scheduled confidence vote), explained that the state of the Indian Government was akin to a superposition of different states, examples being, a "caretaker government" state or a "proper" majority government state.

In his words: "It's a 'probability wave' similar to the cat in the famous thought experiment of Erwin Schrödinger (5-003), where the cat was considered as a wave in a superposition of either a 'live cat' state, or a 'dead cat' state, until an observation was made. As soon as the trust vote is taken, there will be a collapse of the wave function and then and only then will it be clear whether it is a caretaker government state or a proper majority government state. And until we, or anyone else, including the media or the rest of the nation "observe" the results of the proceedings, the superposition of states will continue.

At Cambridge University on the same day, July 21, the following dialogue took place between a certain professor of Indian origin, who is a follower of the Many-Worlds Interpretation of quantum physics, and his students: (5-004)

Professor: If the UPA (United Progressive Alliance, the ruling coalition of political parties of India with Congress as the main party) wins the trust vote then there will also exist a parallel universe where they will lose the trust vote.

Student 1: But, sir that will be the case only if there is a fifty/fifty chance for each, which is not the case as I have just placed a bet with Ladbrokes (the bookmakers) that the UPA will lose the trust vote, and I got odds of 6 to 4 against.

Pr: What percentage chance does that work out as?

St1: That's easy to calculate and is equal to (4 x 100) / (6 + 4) = 40 percent.

Pr: And what were the odds for UPA winning the trust vote?

St1: Six to four on

Pr: Percentage chance?

St1: (6 x 100) / (6 + 4) = 60 percent.

Pr: So, what's the conclusion?

St1: For every six parallel universes where UPA wins the trust vote there will exist four parallel universes where they will lose the trust vote.

Pr: Hey, wait a minute!! The two percentages add up to one hundred. Surely Ladbrokes is not a nonprofit organization.

St 2: I have an explanation for this.

Pr: Please share that with us.

St 2: There is actually a third horse called "dead heat" and as this is a cliffhanger of a race, a dead heat is a possibility, which is given a 10 percent chance by Ladbrokes (they are offering nine to one odds against), and so this 10 percent is their profit margin.

Pr: But how can there be a dead heat when the number of voters is 543, which is an odd number.

St 3: That's possible, as there are expected to be some absentees, and if the number of such absentees is also an odd number then an odd number minus another odd number becomes an even number.

Pr: What then is the final assessment?

St 4: That for every 545 parallel universes where the trust vote is won by UPA there will exist 365 parallel universes where the trust vote is lost by UPA and ninety parallel universes where a dead heat results.

**

Somewhere in Kolkata on the night of July 22, a logician, having just finished reading Erwin Schrödinger's *What is Life?* and realizing that the number of molecules in a glass of water was about ninety times the number of glasses of water that can be taken out of the combined volume of all the oceans, did some more calculations on the volume of the earth's atmosphere and made the following deduction:

"At least one molecule (out of several trillions) breathed out by each and every speaker at the trust vote also happens to be one common molecule (again out of several trillions) breathed out by Julius Caesar when he said 'Et tu Bruté — Then fall Caesar!'"

The next morning on July 23 (one day after the trust vote), after thinking all night and doing some in depth calculation based on the size of the atom and that of its nucleus and realizing that the nucleus while containing all the mass of the atom occupied as little space as a grain of sand in a huge auditorium (5-005), came out with another statement:

"It was an empty hall, even when all the 530-odd members of Parliament were there with all the furniture, tables, and chairs including the air inside, it was indeed an empty hall, empty to such an extent that the quantum of void was nearly fourteen thousand trillion times that of matter."

**

In New Delhi, sometime on the July 23, a certain professor with some idea of time, space, and relativity, while talking to his students, made the following statement (5-006):

"The event corresponding to the announcement of the result of the trust vote by the speaker might have been timed as 2230 hrs (whatever) IST on July 22 by observers sitting in the hall, but for observers watching the proceedings from different points in the universe the event occurred at different times and could not simultaneously be watched by these observers. If one of these observers (let's call him O1) is on a planet E1 orbiting a star S1 located nearly twenty-five light-years away, he will actually be able to see the event when it is the year 2033 on the planet Earth. And, in case one of the members of Parliament, say M1, at the precise time of the announcement, starts traveling at the speed of light toward said planet E1, he will continue to see the same expression on the speaker's face for his entire to-and-fro trip.

Even by the time he returns back from E1, there is no change in the speaker's expression nor in the setup in the Parliament house except that M1 thinks himself to be fifty years older whereas according to the people on Earth they are all relatively older by fifty years than M1. Of course, speed-of-light traveling is an impossibility as M1 will have to be infinitely massive to do so. Nevertheless, even if he travels at less than the speed of light (but sufficiently fast), paradoxical situations would still arise. For example, consider two MPs, M1 and M2, who are twins of age forty years when M1 leaves for planet E1 traveling at 0.5c (where c = speed of light), when M1 returns to the Earth after touching E1, and assuming each of them is watching the clocks fixed in the other's frame the following dialogue is expected to take place.

M2: I am 140 years old but you are only 126 years and seven months old.

M1: I may be 126 years and seven months old but you are only 115 years old.

The fact is that M1 considers length contraction as the parameter for his calculation whereas M2 considers time dilation for his calculation (5-006).

Student 1: Will M1 meet O1 and find him looking at the Earth when he reaches E1?

Professor: No way, speed of light is constant at about three hundred thousand km/sec relative to all observers regardless of the speed at which they travel, and so if M1 is to meet O1 (looking at Earth) he will on reaching E1 have to stay there for twenty-five years.

Student 2: If some observer (say O2) on the planet E1 is looking at the Earth, now what does he see?

PH: Maybe the MPs shaking hands with each other while celebrating the success of the Indian cricket team winning the World Cup. (5-007)

Student 3: It's all too complex; isn't there a simpler definition of relativity?

Professor: One day a pretty young girl asked the same question of Einstein, to which his reply was: "If a gentleman sits next to you and enjoys your company for an hour it will appear like a minute; if he sits on a hot stove for a minute it will appear like an hour — that's relativity."

**

To make matters still more complicated, a cosmologist in Tokyo made the following statement early morning on July 23:

"During the time taken by the speaker to announce the results, he, along with all the MPs in the hall, traveled millions of miles from his original location. For observers in reference frames located in far-off galaxies, the coordinates of the members shifted in a complex manner. While moving along with the planet Earth in an orbit around the sun at nearly seventy thousand miles per hour, they all traveled along with the solar system around a massive black hole located near the center of the Milky Way at even a much faster rate, and the speaker along with the MPs with the planet Earth and the Solar System along with the Milky Way Galaxy traveled away from the said observers at nearly the speed of light."

As for the extra complexity in the motion of the twin M1, I leave it to your imagination.

**

Here in Bangalore a man called 'X' and his friend 'Y' had the following discussion on the evening of the July 23, on being told

81

about the above conversations of the professors and philosophers with their students.

Y: Aren't all these statements and discussions of a trivial nature?

X: (a determinist to the core but an ordinary human being with some knowledge of quantum physics knowing very well that the arrow of time has no meaning at the quantum level and that quantum entities individually move about here and there randomly with complete disregard for the phenomena of cause and effect, but once they combined to form tables and chairs, you and me, or mosquitoes — or even smaller substances, but big enough to be considered within the domain of classical physics — then the phenomena of cause and effect governed their every movement) had this to say in reply to Y's question:

"Perhaps it is true, but sometimes a trivial cause can have a far reaching effect and even a harmless walk in the garden can cause twenty million people to die." (5-001)

Y: Wow that's amazing!!!

But what does all this have to do with the trust vote?

X: It's the same thing, the phenomena of cause and effect. Another walk in the garden by somebody else might have brought someone in this world, who was the grandmother of a man called Einstein. And if it hadn't been for Einstein's special theory of relativity the trust vote of July 22, 2008, would not have taken place at all. In fact, it is possible to work out the precise sequence of events, beginning from Einstein's famous equation $e = mc^2$, and leading up to the trust vote via the nuclear deal, by downloading and studying inputs from the Internet on relevant subjects and views.

In fact, these topics/views broadly highlight the key events that link Einstein's relativity theory to the trust vote, over a period of over one hundred years: understanding of the equivalence

of mass and energy, the properties of uranium, nuclear chain reaction, the finding that such a chain reaction could either be controlled to produce usable energy or allowed to go out of control to
produce a violent explosion, experiments by scientists worldwide
including those of Enrico Fermi and Szilard, potential of nuclear
energy as a source of electric power/nuclear weapons, World War
II, Einstein the follower of Gandhi, Einstein the pacifist and a believer in the One World concept, Einstein the German who detest-
ed the Germans, a certain letter written by Einstein to President
Roosevelt (5-008), the setting up of the Briggs Committee followed
by the National Defense Research Committee, the Pearl Harbor
attack, frantic activity and expanded research at the NDRC, the
Manhattan Project set up for the design and development of the
atomic bomb, bombing of Hiroshima, frantic build up of nuclear
capabilities by powerful nations of the world, the urgency to ensure control of such build up by other nations, the "one world
or none" concept, the NPT, the CTBT, the IAEA and the safe-
guards necessary to ensure that nuclear materials do not go into
the wrong hands, the Atomic Energy Act of 1954, Section 123 of
the act pertaining to cooperation with other nations, the demand-
supply gap in energy, anticipated depletion of certain resources
in the future, the requirement of energy security by nations, the
necessity of more and more nations in the world to partake in the
responsibility to ensure prevention of use of nuclear weapons,
the Indo-US nuclear deal for civilian nuclear cooperation, T\the
Hyde Act being the framework of the Indo-US nuclear deal, the
123 agreement, the Nuclear Suppliers Group, US foreign policy
(sometimes succeeding, sometimes failing), articles written by
the Indian and world media some in favor and some against the
nuclear deal, the interactions responsible for the actions of the
various political groups in India, unprecedented increase in the
rate of increase in oil prices, global inflation including inflation
in India, withdrawal of support by a certain section of the Indian
leadership, the Indian Political system and its flaws (where the
possibility existed that a certain decision "a" can be taken against
a decision "b" even if less than 20 percent are in favor of "a"),
interactions responsible for a certain other group of politicians

to support the ruling group, the confidence and conviction of the Indian Prime minister that he was right, his request for the trust vote, the trust vote.

**

The discussion continued:

Y: That's amazing. What is the implication of this linkage of events?

X: All our actions are caused by interactions of the rest of the world on us including past interactions stored in our brains.

Y: Does it mean all that has happened was destined to happen?

X: "Destiny" and "phenomena of interactions" are two different aspects. With the former, you can explain the past by saying "it could not be helped, it was destined like that," but with the latter, you can explain the past as well as build your own future by interacting to improve your interactions.

**

Now these guys X and Y as well as all the professors and their students, not to forget the two guys who walked in their respective gardens might be imaginary people created in the mind (as a Gedankin or a thought experiment) to explain the interaction phenomena and the weirdness of quantum physics, but the events constituting the sequence from Einstein's relativity to the confidence vote are events that have actually happened and recorded in history. And, in fact, if the part of the write up (pertaining to the sequence in question) is enlarged to a twenty-page document, you will find that one of these events in the sequence was in actual fact a walk in the snow-covered woods outside Stockholm, a walk that was to shape the future of the human race (5-009). It was a walk during which the walkers, the scientists Lise Meitner and her nephew Otto Frisch, discussed the findings of an experimental research carried out by another scientist called Otto Hahn, and during discussions they discerned something of extreme significance (relating to the splitting of the uranium nucleus and consequent release of energy), and this vital piece of information was promptly passed on by telephone to the famous scientist Niels Bohr, who, at that time, was on his way to an important conference, leading ultimately to the experiments undertaken on war footing by Szilard and Fermi (mentioned in the sequence). The complete real-life drama of what transpired in the late 1930s and early 1940s in different parts of the world involving scientists and political leaders of several countries as the characters, and uranium as the element in demand to be grabbed at all costs from wherever it was being mined at the time is best described in Ronald W. Clarke's *Einstein the Life and Times*, surpassing the best of thrillers in intrigue and drama. Here, it is not to be inferred that there would have been no nuclear devices had some of these events not taken place, but the possibility did exist that if one or more of these events had been delayed or not taken place at all, some other country, possibly Germany, may have made first use of the bomb, thus altering history and bringing in completely different kinds of interactions with no such confidence vote of July 22, 2008.

**

Concluding session held at Bangalore on July 25, 2008:

Y: Is the world moving in the right direction?

X: Perhaps, but we have to be careful. There is an analysis done by a certain scientist named Ed Harrison, in his book *Cosmology*, according to which the technological civilization of the planet Earth may have reached the top hundred in the Milky Way, but it may not reach the top ten if it self-destructs before that (5-010).

Y: How can we reach the top ten in the Milky Way?

X: By adopting a global approach on the assumption that a typical human being, anywhere in the world wants to live in peace, and the average moral standard in any country is the same for all countries and does not differ from country to country as it is not a measure of "temperature" or "humidity." If this coefficient, corresponding to the average moral standard is denoted by "qx," and if the corresponding coefficient pertaining to the nation's leaders is denoted by "qy", then it must be ensured that qy exceeds qx, and one by one all nations must become members of a club of nations whose qy exceeds qx and the primary role of these club members is to see that those nations who are not yet members of this club should become members. Of course, it goes without saying that the designated representatives of all member countries should address the important issues of the planet, the foremost being the control on the use of nuclear weapons. (5-011).

Y: Has India a role to play in the planet's escape from self-destruction.

X: All nations have to play their part and with India's emergence as a fast-growing economy and its newfound stature as a responsible and trustworthy nation, not to mention its neutral stance and friendly approach, it is looked upon by all nations big or small as a welcome addition to the elite group of nations entrusted with the responsibility toward the future security of the world not just in preventing wrong hands from getting powerful but also in preventing powerful hands from possible wrong actions.

**

STATUS UPDATE: SEPTEMBER 2009:

The ruling party returned to power with a substantially improved majority. Parties with negative ideologies suffered major setbacks. They must learn from their mistakes. The voting public became much wiser after the confidence vote. And they will get wiser and wiser as time goes by. Einstein did nuke the Indian political system.

SEPTEMBER 2013:

There is a virus called "corruption at high places," which has infested the ruling party. What will the voting public do? The future is uncertain to say the least…not just of India…but of the whole world.

CONCLUDING PARAGRAPH:

"There has been one and only one sequence of events from the Big Bang through the entire history of the universe with all its galaxies, star formations, supernovas, planetary systems, including our own Solar System, Earth, evolution of life on Earth, the human race, your arrival in this world and immediate containment in a speck of jelly, your growing up, continuous interactions of the rest of the world on you, making you precisely what you are up to the present moment leading you to read this article. There was no way you could have avoided reading it, and now is the time to act and play your part to ensure peace and prosperity for your nation and for the world."

Back to relativity, the equation $e = mc^2$ should be used for the development of the world and not for its destruction.

**

The realization must manifest itself that probabilities need to be created for such actions that will ensure that the collapse of civilization is prevented. A *sequence* of events needs to be designed…and implemented…that can lead…toward a "seminar" that can prevent the extinction…of the human race.

ENDNOTES

5-001:

One day late in the evening, just before going to bed, A man called Z takes a five-minute walk in the garden. One-hundred-fifty years later nearly twenty million people die. They wouldn't have died had Z not taken that walk in the garden. How is that possible? Any guesses?

He was bitten by a mosquito...a few days later he was down with malaria...was treated in a hospital...Came back home after the treatment. One-hundred-fifty years later nearly twenty million people die. They wouldn't have died had Z not taken that treatment in the hospital. How is that possible? Any guesses?

In the hospital, he was cared for by a nurse, they fell in love, they later got married, and they had a daughter. So? She was Hitler's grandmother.

5-002:

COPENHAGEN INTERPRETATION:

The collection of ideas—uncertainty, complimentarity, probability, and the disturbance of the system being observed by the observer—is together referred to as the Copenhagen interpretation of quantum mechanics. The strangest thing about the standard CI of the quantum world is that it is the act of observing a system that forces it to select one of its options, which then becomes real. At the heart of CI. is the phenomena called "the collapse of the wave function" (John Gribbin).

"collapse of the wave function = reduction of the state vector"

"And this is used to describe the procedure that is adopted when an effect is magnified from the quantum to the classical level" (Roger Penrose).

"Essence of the Copenhagen interpretation: The world must be actually observed to be objective" (H.R.Pagels).

Contrary view: "CI and objective reality are incompatible" (Einstein).

"State vector (SV): The mathematics expression describing one of two or more states that a quantum system can be in, for instance an electron can be in either of two spin states, called spin up and spin down. The amusing thing about quantum mechanics is that each SV can be regarded as the superposition of other state vectors" (R.A Wilson).

Extreme view: If no one is looking at the moon, it doesn't exist.

As explained in chapter four, the nonexistence of objective reality until measurement takes place is one of the key elements of the Copenhagen interpretation of quantum mechanics. The EPR paper intended to disprove this by explaining that if this was true, then there exist nonlocal actions in this universe. The EPR paper could not disprove it, instead, Bell's inequality and the Aspect experiment showed nonlocal action as another feature of quantum physics. The God that plays dice played a trick on us by permitting quantum entanglement without giving us the power of sending any useful information at a speed faster than light. It appears that the key objection to this key element of CI (nonexistence of objective reality) has been ruled out and supporters of CI are therefore persisting with it.

Whether it is true or not true, all I can say is that this objection to objective reality, can be culturally very damaging to the common man's beliefs, and for the time being, this concept should be left alone, and need not be part of any philosophical discourse or debate.

Once our technological civilization scrapes through the next three hundred years, and its collapse is averted, and a suitable stage is reached (perhaps thousands of years hence), when the ratio of super intelligent beings over less intelligent beings far exceeds unity, we can reopen the debate and come to a suitable consensus.

5-003:

SCHRÖDINGER'S CAT:

A thought experiment devised by Erwin Schrödinger — in its simplest form it comprises a box or a chamber containing a cat and a certain device programmed in such a way that there is a precise 50 percent probability that a certain substance is emitted by the device just once during a fixed time interval. If emitted, it converts the air inside the chamber into a poisonous gas that kills the cat. An observer called X then opens the box and looks inside to see if the cat is dead or alive. As per the standard Copenhagen interpretation of quantum mechanics, the observer is not just an observer but a participator in the experiment. Prior to the observation, the cat is in a state of superposition (a probability wave) of a live cat as well as a dead cat. The act of observation causes a collapse of the wave function and the observer sees either a dead cat or a live cat.

5-004:

MANY WORLDS INTERPRETATION:

In this interpretation, initiated by Hugh Everett, wave functions never collapse and every conceivable outcome embodied in the wave function materializes albeit in different parallel universes. Whenever the world is faced with a choice at the quantum level, the universe divides itself into as many parts as there are choices so that all possible options are followed. Thus, in the Schrödinger's cat experiment, if X finds the cat dead, then there exists a parallel universe where X finds the cat to be alive and kicking.

And of course there are several parallel universes where you have not been reading this sentence.

If this parallel universe concept is to be believed, then we must also assume that there exist some parallel universes, where this concept is not believed in, and in some cases not even heard of. A majority of parallel universes must be assigning a certain probability (varying from almost negligible to almost certain) to the existence of parallel universes. Almost all of those will be virtually certain of the impossibility of jumping from one parallel universe to another. But, in those parallel universes where the concept of parallel universes has been proved by experiment, then going from one universe to another may not be impossible. But in all probability the God that plays dice may have played some tricks and may not permit such inter-universe travel, in the same way as He is preventing us from passing on useful information at a speed faster than light. Nevertheless, this concept of the Many Worlds Interpretation carries an incredible quotient of metaphysical baggage and has little chance of success of being accepted by the current intelligence level prevailing on the planet Earth.

However the MWI may not, like the CI, cause any sort of cultural damage to the common man's conscience. They are likely to find it quite amusing, and there is no harm if science fiction writers are able to entertain people through this novel idea. I am, in fact, myself trying to make use of it by composing a seminar held in a parallel universe, where it was shown live on television and proved successful in preventing the collapse of civilization, and incorporating it in this autobiography, in the hope that some day it will lead to an actual seminar.

Notwithstanding the quotient of metaphysical baggage, the Many Worlds Interpretation along with the Copenhagen interpretation are the two front runners among the many interpretations of quantum mechanics. The last paragraph of Footnote 5-002 is applicable here too:

Once our technological civilization scrapes through the next three hundred years, and its collapse is averted, and a suitable stage is reached (perhaps thousands of years hence), when the ratio of super intelligent beings over less intelligent beings far exceeds unity, we can reopen the debate and come to a suitable consensus. However, this subject will be discussed extensively in chapter six — "Six Words — Seminar Held in a Parallel Universe."

5-005:

This can be worked out by considering that:
    The radius of an atom is about 0.000000005 cm.
    And the radius of the nucleus of an atom is about 0.00000000000005 cm

5-006:

All these are consequences of Einstein's special relativity theory governing physical phenomena when one object or reference frame moves relative to another at a speed comparable to that of light, i.e., 300,000 km/sec.
    Mass Increase: The mass of a body increases with velocity. In Einstein's formula, M is given by:

$$M = Mo/(1 - v^2 / c^2)^{0.5}$$

    Where Mo is the rest mass of the body, v is the speed at which it moves and c is the speed of light.
    Length Contraction: The length of an object in a frame through which it moves determined by viewing its ends at the same time (two events) in that frame, is smaller than the length of the object in the frame in which it is at rest.

$$L = Lo / Yv$$

    Where L = length of an object in the frame relative to which the object moves.

Lo = Length of the object in the frame in which it is at rest.

And $Yv = 1 / (1 - v^2 / c^2)^{0.5}$

    Time Dilation: Two events occurring in the same location in one frame will be separated by a longer time interval in a frame moving relative to the first.

$Tt = Yv*To$

The time interval Tt that passes in another frame is longer by the factor Yv than the interval To that passes in the frame in which events all occur at the same location.

Relative Simultaneity: Two events at different locations simultaneous in one frame of reference will not be simultaneous in a frame of reference moving relative to the first.

Fundamental postulates of special relativity: "The form of each physical law is the same in all inertial frames. (By definition an inertial frame is a frame in which an object experiencing zero net force, i.e., a "free object," moves at constant velocity)."

Light moves at the same speed relative to all observers.

For further reading: *Nonclassical Physics* by Randy Harris, *Six Not So Easy Pieces* by Richard P. Feynman.

Important note: The subjects of "special relativity" "speed of light as the limiting speed," "length contraction," "quantum entanglement," and "the interpretation of quantum mechanics including a certain "transactional interpretation" are inextricably linked and have far reaching philosophical impactions in the derivation of principles, which are acceptable to one and all. On the basis of these principles, all the religions of the world can be combined to form a common religion. This is the central theme of the book as well as of the next chapter on "The Seminar."

5-007: This World Cup victory took place in 1983.

5-008:

Two letters written by Einstein to President Roosevelt in August 1939 and March 1940—the latter actually written to Dr. Sachs for onward transmission to Roosevelt—warning of the German research in fission and urging the President to initiate and then speed up a research program in the United States to explore the feasibility of atomic bombs, were considered by many as the biggest and most powerful interactions between Einstein and the United States. The German occupation of Czechoslovakia had brought a halt to the sale of uranium and evidence pointed to intensive nuclear research by German scientists. The two letters along with a third letter that resulted in the Manhattan Project — contents reproduced below — contributed significantly toward enormous atomic research effort in the United States that ultimately tilted the scales in favor of the United States in the race toward the first use of the atomic bomb.

Contents of the third letter (in part):

"I am convinced as to the wisdom and the urgency of creating the conditions under which that and related work (referring to the research work of Szilard and Wigner) can be carried out with greater speed and on a larger scale than hitherto. I was interested in a suggestion made by Dr. Sachs that the Special Advisory Committee supply names of persons to serve as a board of trustees for a nonprofit organization, which, with the approval of the government committee, could secure from governmental or private sources or both the necessary funds for carrying out the work. Given such a framework and the necessary funds, it (the large scale experiments and exploration of practical applications) could be carried out much faster than through a loose cooperation of university laboratories and government departments."

The nonprofit organization corresponds of course to the organization responsible for carrying out the Manhattan Project. The letter was implemented briskly. The "Briggs Committee" was drastically reorganized and brought under the wing of the National Defense Research committee created by Roosevelt, and a special committee of the National Academy of Sciences set up to inform the government of any development of nuclear fission that might affect defense.

Though Einstein was an out-and-out pacifist, there was no way he could have not written these letters as the consequence of a German first-use of atomic bombs might have been catastrophic.

Had these letters not been written, we have no idea what would have been the history of the planet Earth for the last about seventy years.

For further reading:

Einstein – The Life and Times by Ronald W. Clarke

Some Strangeness in the Proportion (chapter on "Albert Einstein — Encounter with America") edited by Harry Woolf

5-009:

Discovery of Fission

In the words of Otto Frisch (recalled in his memoirs):

"When I came out of my hotel room after my first night in Kungalve, I found Lise Meitner studying a letter from Hahn and obviously very puzzled by it. I wanted to discuss with her a new experiment that I was planning, but she wouldn't listen: I had to read that letter. Its content was so startling that I was at first inclined to be skeptical. Hahn and Stresemann had found that those three substances were not radium…(but) barium.

The suggestion that they might after all have made a mistake was waved aside by Lise Meitner: Hahn was too good a chemist for that, she assured me…

"We walked up and down in the snow, I on skis and she on foot (she said and proved that she could get along just as fast that way), and gradually the idea took shape that this was no chipping or cracking of the nucleus but rather a process to be explained by Bohr's idea that the nucleus is like a liquid drop; such a drop might elongate and divide itself…We knew that there were strong forces that would resist such a process, just as the surface tension of an ordinary liquid drop resists its division into two smaller ones. But nuclei differed from ordinary drops in one important way: they were electrically charged, and this was known to diminish the effect of the surface tension.

At this point we both sat down on a tree trunk and started to calculate on scraps of paper. The charge of the uranium nucleus we found was indeed large enough to destroy the effect of surface tension almost completely, so the uranium nucleus might be a very wobbly, unstoppable drop, ready to divide itself at the slightest provocation (such as the impact of a neutron). But, there was another problem. When the two drops separated, they would be driven apart by their mutual electric repulsion and would acquire a very large energy, about 200 MeV in all; where could that energy come from? Fortunately, Lise Meitner remembered how to compute the masses of nuclei from the so-called packing fraction formula, and in that way she worked out that the two nuclei formed by the division of a uranium nucleus would be higher than the original uranium nucleus by about one-fifth of the mass of a proton. Now whenever mass disappears, energy is created, according to Einstein's formula $e = mc^2$, and one fifth of a proton mass was just equivalent to 200 MeV. So, here was the source for that energy; it all fitted."

For further reading:

Lise Meitner – A life in Physics by Ruth Lewin Sime

The work of a genius…a masterstroke…ushering in unlimited prospects toward achievement of energy security for the world… all in very good taste…but, of course, perhaps a step closer toward the "collapse of civilization."

5-010:

The number of stars in the Milky Way is so large that the probability of at least one hundred technological civilizations at least as intelligent as the one on the planet Earth is reasonably high as worked out below:

(Refer to *Cosmology* by Edward R. Harrison—the chapter on "Life in the universe")

If N = Number of stars in the galaxy (Milky Way),

Then the number of Technological Civilizations (TCs) that may have existed/still existing is given by:

n = N * P1 * P2 * P3 * P4 * P5 * P6 * P7*

Where P1 is the fraction of stars in the galaxy similar to the Sun, i.e., not too blue, not too red and not members of closed binary systems. Reasonable estimate one-tenth, i.e., 0.1.

P2 is the fraction of sunlike stars having earthlike planets. Optimistic estimate 1.0, and conservative estimate 0.1.

P3 is the fraction of such planets occupying a habitable zone, i.e., neither too close (like Venus), nor too far (like Mars) from the parent star. Reasonable estimate 0.1.

P4 is the probability that life originates in a unicellular form. Reasonable estimate 0.1.

P5 is the probability that it evolves into complex multicellular organisms like mammals. Many optimists may consider P5 to be close to unity but recognizing the hazards and traps involved, a conservative estimate is 0.1.

P6 is the probability that life develops intelligence. Again, many optimists consider this inevitable but ERH again takes a conservative approach. He feels that the environment must in some way administer rude shocks in the right way at the right times to the right species, so that natural selection will favor the development of large brains and the probability of this happening is very small — about 0.1.

P7 is the probability that intelligent life discovers science and develops an advanced technological civilization. Here again an optimist would take P7 as almost unity but a pessimist would take a different view. Science was not discovered by the cultures of Africa, America, China, India, and Japan, etc. Science arose because of accidental and improbable circumstances that existed in Greece and later in Europe. Efforts of a few individuals, who rejected the magic of gods, can be considered as the first steps that created science. Later it was revived in Europe in the face of organized hostility. Again, a conservative estimate of 0.1 is assumed for P7.

Based on above-estimated values of the factors P1 to P7, the likely number of TCs (in the Milky Way only and not the universe as a whole) that may have existed sometime in the past, inclusive of those that are still existing, varies from about ten million (optimistic estimate) to about ten thousand (pessimistic estimate). According to ERH, this number is closer to the lower estimate.

How many of these TCs still exist today?

If this number is denoted by n1, its value is equal to P8*n. What is P8?

There are many, many ways in which a TC can collapse, such as cosmological collisions (asteroid hits in case of Earth), climatic and other conditions no more suitable for continuance of life, but none of these has a greater probability than self-destruction by aggressive approach. Developement of science creates new hazards and the chances are that most TCs are short lived. Conflicts, wars, terrorism, misadventure, and above all the possibility of nuclear wars, gives a low value of 0.1 (optimistic) or 0.01(pessimistic) to the factor P8.

And so out of ten thousand TCs that have arisen in the galaxy, perhaps between one hundred and one thousand are still surviving.

How many of these TCs, which are still surviving, are likely to survive permanently?

If this number is denoted by n2, its value is equal to P9*n1. What is P9?

Two objectives must be fulfilled in order to survive permanently (at least as long as the Stelliferrous Age):

Galactic Colonization: The TC must survive about fifty million years (by avoiding self-destruction) to achieve this objective, to develop interstellar space travel, to develop fusion power and other technologies to construct large space vehicles that can travel at one-thousandth the speed of light, as well as have its own biosphere containing a social unit of tens of thousands of people to enable ten thousand years of travel time at a stretch, by halting at one destination for ten thousand years before embarking on the next journey. In this way it is possible to diffuse outward at ten light-years per twenty thousand years and colonize a substantial portion of the galaxy in fifty million years (which is not even 0.5 percent of the present age of the universe).

Galactic Selection: The TC must understand and implement properly a certain law called "The Biogalactic Law." Given below is a very interesting paragraph from ERH's book referred above, on this subject:

"The Biogalactic Law: We are the outcome of natural selection and to the operation of this law can be attributed the fitness of the human body and brain. Presumably this is true also of the life on other planets that attains a state of advanced intelligence. When a civilization gains control of its planetary environment, the evolutionary game changes, however, and new rules determine what is fit and unfit. Natural selection now operates on a planet-wide scale according to a biogalactic law that will be referred to as galactic selection. This speculative law of galactic selection states simply: "Intelligent life forms that are destructively aggressive do not colonize the galaxy. This law operates conceivably in two modes; the first is unconscious and automatic, and the second is conscious and deliberate."

The two words "do not" in the above definition of the biogalactic law probably correspond to the five words "will not be permitted to." It goes without saying that the above objectives are indeed very difficult to achieve and even a very optimistic value that can be accorded to the factor P9 cannot exceed 0.1. Hence the conclusion: The Technological civilization of the planet Earth may have reached the top hundred in the Milky Way, but it may not reach the top ten if it self-destructs before that.

5-011:

The current state of the technological civilization on the planet Earth (with the enormous content of nuclear arsenals at its disposal) is comparable to a fast-moving vehicle traveling at high altitude in a hilly terrain having to negotiate innumerable zigzags. At one stage (near the end of World War II sometime after the discovery of fission) the vehicle was on the verge of overturning, but was prevented by expert handling and deft maneuvering by the driver, a minor accident resulted but the vehicle was brought on course (Einstein's letters to Roosevelt—expert handling and deft maneuvering—leading to the bombing of Hiroshima—a minor accident—but avoiding a German first use of a nuclear device). The vehicle started moving erratically again (the Cold War years of the 1960s) but managed to return back on course. (A German first use of the nuclear bomb would have been catastrophic for the world, which could be considered a major accident – like overturning of the vehicle -. Compared to that the bombing of Hiiroshima, which ended the war, can be considered as a minor accident. The Cuban missile case is another example of a narrow escape from a catastrophic situation (overturning). The subject is discussed in great detail in chapter SixE in the Q and A session of the seminar.) In recent years the speed of the vehicle and the risk of overturning have increased further and the driver (the leaders of nations) has enormous responsibility to keep the vehicle in control. The next three hundred years are considered the most dangerous zone and must be maneuvered as smoothly as possible if it is desired to improve the probability factor P9.

Chapter Six

THE SEMINAR

**

CHAPTER SIX A

THE ARRANGEMENTS

"There is someone in my head but it's not me." *The Dark Side of the Moon* by Pink Floyd

It maybe that "someone" is something—such as a mind—offloaded from a UFO. (Refer 6 -013 in SIX B)

And that someone or something is currently observing what's going on here. And I tell you its not good. Unless there is a sea change in attitudes. Conflicts...wars...population explosion, etc...five hundred years looks like an upper limit before extinction time...

**

In a certain parallel universe, there is a company called "Revised Greats Incorporated," located in Interlaken, Switzerland. This company creates (by selection and comprehensive training) revised versions of great scientists and philosophers of the past. These revised versions are mostly selected from retired Hollywood or Bollywood actors and are required to act in a way, the great people — whose revised versions they are — would have acted were they still alive and fully aware of the events of the past, including the events that took place on the planet Earth since the death of the original ones. The company then sends these great people (the revised ones) as consultants to various meetings, conferences, seminars, etc., wherever required. Profit, of course, is one of the motives. In some cases, such as Einstein for example, there are several revisions (R1, R2, R3, etc.) available for deployment...a question of demand and supply.

In the said parallel universe, where the human civilization of the planet Earth was as precariously placed — with regard to the probability of its self-destruction — as in this universe, three bright and enterprising ladies organized a seminar at the Institute of Advanced studies, Bangalore, in June 2014. The planning of the seminar was started six months in advance. There were twenty-five speakers invited to give their talks, mostly selected from the "Revised Greats Inc."

The seminar with the strange title "Six Words" was shown live on TV channels throughout the world. Philosophical statements from which the six words could be decoded were displayed on the screen in a continuous manner.

There were 160 registered delegates from all walks of life. All the delegates were by invitation only. They were all rational thinkers. They were all fully aware of the current state of affairs on the planet earth...the rising technological advances and the dangers posed by the rising technology...the rising population of the world and the dangers posed by the rising population...the conflicts between nations and the dangers posed by the nuclear arsenal available with some of these nations in conflict. They all were required to have an unbiased approach, as if they were men from outer space with no special attachment with any country or any religion. They all understood their responsibilities toward ensuring that the "desired objectives" are achieved. And yours truly was privileged to be one of these invited delegates.

And "yours truly" was asked — by one of the three organizers — to be prepared to come on stage on the third morning of the seminar...for a good three to four hours...and enlighten the audience on the "six words."

Why me? What do I know? I am just a structural engineer.

"Thats why," said the lady. The technological civilization of the planet Earth is like a tall building that has developed cracks due to overloading and differential settlement of the foundations; the collapse is imminent; it's just a question of time before the cracks start widening and failure starts.

Stephen Hawking thinks two hundred years is all the time left for us. Martin Rees is even more pessimistic and feels the twenty-first century is our "final century." They all know the most likely causes: rising populations, depletion of resources, and nuclear wars. Stephen Hawking has even suggested that colonizing the Galaxy is the only alternative and must be achieved within these two hundred years, which, of course, appears to be out of the question.

You will be required to redesign the structure.

I understood what she meant. Here, the "overloading" corresponds, of course, to the population explosion; the "differential settlement of foundations" corresponds to the nonuniform depletion of resources that support the structure; the "development of cracks" corresponds to the phenomena of conflicts and wars; and "collapse and failure" correspond to nuclear warfare and its chain reactions. Stephen Hawking thinks we have no option other than vacating the premises and shifting to other buildings. But where is the new land? And where is the time to search for this new land and then design and construct new buildings?

How much time is required to develop fusion power and other technologies to construct large space vehicles that can travel at least as fast as one thousandth the speed of light? Even at that speed it will take ten thousand years to cover about ten light-years' distance from one planetary system to another. After halting at these destinations for about ten thousand years, we may then move forward onto the next interstellar journeys. The space vehicles have to be large enough to have their own biosphere, containing a social unit of millions of people. These objectives are not impossible to achieve, but even a highly optimistic assessment would require a few thousands of years; it certainly looks improbable in a few hundred years, which is the time left at our disposal.

I was in dire straits...stressed to the limit...not just tensile... but even twisting moment. What could I do? There was no way out...I must concentrate fully on what goes on during the first two days...and I hoped and prayed for that "someone" to remain forever in my head.

The seminar was scheduled to last for four days. One of the three ladies was the master (or the mistress) of ceremonies hereinafter denoted by the symbol "MOC." She outlined the agenda as follows:

During the first two days and a portion of the third day the speakers were to dwell on the topics of "the philosophy of science and religion and the establishment of principles on which the human beings must order their lives in order that the human civilization can survive the next few hundred years considered as 'the danger zone,' and thereafter live in peace for a sufficient length of time to plan and initiate action toward the 'colonization of the galaxy.'"

The topics for the remaining part of the third day would be related to the "problems faced by the planet Earth that could lead to a very high probability of the collapse of civilization within the next three hundred years, an understanding of the current status of the world affairs and the nature of conflicts between nations, the interactions of the world responsible for the current state, and the measures required at broad levels to resolve the problems."

The agenda for the fourth day would be made known on the fourth day.

**

Chapter Six B

LEADING SCIENTISTS AND PHILOSOPHERS VIEWS ON SCIENCE AND PHILOSOPHY

After outlining the schedule, MOC introduced the twenty-five men on the dais, and requested the one in the center — who was also the Chief Guest — to declare the seminar open and to give the keynote address to the audience.

This gentleman was none other than the revised version of Albert Einstein hereinafter called "Einstein R01" and denoted by the symbol "E."

After the usual formalities the gentleman called "E" addressed the gathering; he spoke for an hour and forty minutes. Given below is an extract from his speech.

COSMIC RELIGIOUS FEELING

(Extract from "Ideas and Opinions" by Albert Einstein)

"Everything that the human race has done and thought is concerned with the satisfaction of deeply felt needs and the assuagement of pain. One has to keep this constantly in mind if one wishes to understand spiritual movements and their development. Feeling and longing are the motive force behind all human endeavor and human creation, in however exalted a guise the latter may present themselves to us. Now, what are the feelings and needs that have led men to religious thought and belief in the widest sense of the words? A little consideration will suffice to show us that the most varying emotions preside over the birth of religious thought and experience. With primitive man it is above all fear that evokes religious notions—fear of hunger, wild beasts, sickness, and death. Since at this stage of existence understanding of causal connections is usually poorly developed, the human mind creates illusory beings more or less analogous to itself on whose wills and actions these fearful happenings depend. Thus one tries to secure the favor of these beings by carrying out actions and offering sacrifices, which, according to tradition handed down from generation to generation, propitiate them or make them well disposed toward a mortal. In this sense, I am speaking of a religion of fear. This, though not created, is in an important degree stabilized by the formation of a special priestly caste, which sets itself up as a mediator between the people and the beings they fear, and erects hegemony on this basis. In many cases, a leader or ruler or a privileged class whose position rests on other factors, combines priestly functions with its secular authority in order to make the latter more secure; or the political rulers and the priestly caste make common cause in their own interests.

"The social impulses are another source of crystallization of religion. Fathers and mothers and the leaders of larger human communities are mortal and fallible. The desire for guidance, love, and support prompts men to form the social or moral conception of God. This is the God of Providence, who protects, disposes, rewards, and punishes; the God who, according to the limits of the believer's outlook, loves and cherishes the life of the tribe or of the human race, or even life itself; the comforter in sorrow and unsatisfied longing; he who preserves the souls of the dead. This is the social or moral conception of God. The development from a religion of fear to moral religion is a great step in people's lives. And yet that primitive religions are based entirely on fear and the religions of civilized people purely on morality is a prejudice against, which we must be on guard. The truth is that all religions are a varying blend of both types, with this differentiation: that on the higher levels of social life the religion of morality predominates.

Common to all these types is the anthropomorphic (6-001) character of their conception of God. In general, only individuals of exceptional endowments and exceptionally high-minded communities, rise to any considerable extent above this level. But there is a third stage of religious experience, which belongs to all of them, even though it is rarely found in the pure form: I shall call it "cosmic religious feeling." It is very difficult to elucidate this feeling to anyone who is entirely without it, especially as there is no anthropomorphic conception of God corresponding to it. The individual feels the futility of human desires and aims and the sublimity and marvelous order, which reveal them both in nature and in a world of thought. Individual existence impresses him as a sort of prison and he wants to experience the universe as a single significant whole. The religious geniuses of all ages have been distinguished by this kind of religious feeling, which knows no dogma and no God conceived in man's image.

Question from the audience: But how can cosmic religious feeling be communicated from one person to another if it can give rise to no definite notion of a God and no theology?

E: In my view, it is the most important function of art and science to awaken this feeling and keep it alive in those who are receptive to it.

We thus arrive at a conception of the relation of science to religion very different from the usual one. When one views the matter historically, one is inclined to look upon science and religion as irreconcilable antagonists, and for a very obvious reason. The man who is thoroughly convinced of the universal operation of the law of causation cannot for a moment entertain the idea of a being who interferes in the course of events — provided, of course that he takes the hypothesis of causality really seriously. He has no use for the religion of fear and equally little for social or moral religion. A God who rewards and punishes is inconceivable to him for the simple reason that a man's actions are determined by necessity, external and internal, so that in God's eyes he cannot be responsible, any more than an inanimate object is responsible for the motions it undergoes. Nobody, certainly, will deny that the idea of the existence of an omnipotent, just, and omnibenificient personal God is able to accord man solace, help, and guidance; also, by virtue of its simplicity, it is accessible to the most undeveloped mind. But, on the other hand, there are decisive weaknesses attached to the idea itself, which have been painfully felt since the beginning of history. That is, if this being is omnipotent, then every occurrence — including every action, every human thought, and every human feeling and aspiration — is also his work; how is it possible to think of holding men responsible for their deeds and thoughts before such an almighty being? In giving out punishment and rewards, he would, to a certain extent, be passing judgment on himself. How can this be combined with the goodness and righteousness ascribed to him?

Science has, therefore, been charged with undermining morality, but the charge is unjust. A man's ethical behavior should be based effectually on sympathy, education, and social ties and needs; no religious basis is necessary. Man would indeed be in a poor way if he had to be restrained by fear of punishment and hope of reward after death.

It is, therefore, easy to see why the churches have always fought science and persecuted its devotees. On the other hand, I maintain that the cosmic religious feeling is the strongest and noblest motive for scientific research. Only those who realize the immense efforts and, above all, the devotion without which pioneer work in theoretical science cannot be achieved are able to grasp the strength of the emotion out of which alone such work, remote as it is from the immediate realities of life, can issue. What a deep conviction of the rationality of the universe and what a yearning to understand, were it but a feeble reflection of the mind revealed in this world, Kepler and Newton must have had to enable them to spend years of solitary labor in disentangling the principles of celestial mechanics! Those whose acquaintance with scientific research is derived chiefly from its practical results easily develop a completely false notion of the mentality of the men who, surrounded by a skeptical world, have shown the way to kindred spirits scattered widely through the world and the centuries. Only one who has devoted his life to similar ends can have a vivid realization of what has inspired these men and given them the strength to remain true to their purpose in spite of countless failures. It is cosmic religious feeling that gives a man such strength. A contemporary has said, not unjustly that in this materialistic age of ours the serious workers are the only profoundly religious people."

...Thunderous applause...

There was a tea break for 15 minutes.

The next speaker was a gentleman who acted as a combination of revised versions of two great scientists: Erwin Schrödinger and Charles Sherrington (having studied their lives extensively) hereinafter denoted by the symbol "SS."

SS spoke for about an hour on "Free will and determinism" in the pre-lunch session, and for about an hour and twenty minutes in the post-lunch session.

Here is an extract from his post-lunch talk:

THE ARITHMETICAL PARADOX

THE ONENESS OF MIND

(Extracts from *What Is Life?* by Erwin Schrödinger)

"The reason why our sentient, percipient, and thinking ego is met nowhere within our scientific world picture can easily be indicated in seven words: because it is indeed that world picture. It is identical with the whole and therefore cannot be contained in it as a part of it. But, of course, here we knock against the arithmetical paradox; there appears to be a great multitude of these conscious egos, the world however only is one. This comes from the fashion in which the world-concept produces itself. The several domains of "private" consciousness partly overlap. The region common to all where they all overlap is the construct of the "real world around us." With all that an uncomfortable feeling remains, prompting such questions as: Is my world really the same as yours? Is there one real world to be distinguished from its pictures introjected by way of perception into every one of us? And, if so, are these pictures like unto the real world, or is the latter, the world "in itself," perhaps very different from the one we perceive?

Such questions are ingenious, but in my opinion very apt to confuse the issue. They have no adequate answers. They all are or lead to, antinomies (6-002) springing from the one source, which I called the arithmetical paradox; the many conscious egos from whose mental experiences the *one* world is concocted. There is only one possible solution to this paradox of numbers that answers these questions appropriately, namely the unification of minds or consciousnesses. There multiplicity is only apparent, in truth there is only one mind. This is the doctrine of the Upanishads. And not only of the Upanishads. Let me quote as an example outside the Upanishads. An Islamic Persian mystic of the thirteenth century, Aziz Nasafi (translated into English from a German translation in a paper by Fritz Meyer):

"On the death of any living creature, the spirit returns to the spiritual world, the body to the bodily world. In this however only the bodies are subject to change. The spiritual world is one single spirit who stands like unto a light behind the bodily world and who, when any single creature comes into being, shines through it as through a window. According to the kind and size of the window less or more light enters the world. The light itself however remains unchanged."

If you refer to Aldous Huxley's *The Perennial Philosophy*, which is an anthology from the mystics of the most various periods and the most various peoples, you will find many beautiful utterances of a similar kind. You are stuck by the miraculous agreement between humans of different race, different religion, knowing nothing about each other's existence separated by centuries and millennia, and by the greatest distances that there are on our globe.

Another member from the dais, a representative of "Amaury de Riencourt" intervened in concurrence with SS: "This ancient wisdom is not limited to the Upanishads and Vedanta. The Chinese Taoists long ago expressed the same monistic viewpoint; the *Tao Te Ching* states that "Therefore the sage embraces the oneness (of the universe), making it his testing-instrument for everything under Heaven." Or we read in the Kuan Tzu book: "Only the chun-tzu (gentleman) holding to the idea of the One can bring about changes in things and affairs."

SS continued: "Thank you R. Still, it must be said that to Western thought this doctrine has little appeal, it is unpalatable, it is dubbed fantastic, unscientific. Well, so it is because our science — Greek science — is based on objectivation, whereby it has cut itself off from an adequate understanding of the subject of cognizance, of the mind. But I do believe that this is precisely the point where our present way of thinking does need to be amended, perhaps by a bit of blood — transfusion from Eastern thought. That will not be easy, we must beware of blunders — blood transfusion always needs great precaution to prevent clotting. We do not wish to lose the logical precision that our scientific thought has reached, and that is unparalleled anywhere at any epoch.

Still, one thing can be claimed in favor of the mystical teaching of the "identity" of all minds with each other and with the supreme mind: The doctrine of identity can claim that it is clinched by the empirical fact that *consciousness is never experienced in the plural, only in the singular.* Not only has none of us ever experienced more than one consciousness, but there is also no trace of circumstantial evidence of this ever happening anywhere in the world. If I say that there cannot be more than one consciousness in the same mind, this seems a blunt tautology — we are quite unable to imagine the contrary — i.e., the plurality of consciousnesses in one mind. We can pronounce these words all right, but they are not the description of any thinkable experience. Even in the pathological cases of a 'split personality' the two persons alternate; nay this is just the characteristic feature that they know nothing about each other.

"I find it utterly impossible to form an idea about either how, for example, my own conscious mind (that I feel to be one) should have originated by integration of the consciousnesses of the cells (or some of them) that form my body, or how it should at every moment of life be, as it were, their resultant. One would think that such a "commonwealth of cells" as each of us is would be the occasion *par excellence* for mind to exhibit plurality if it were at all able to do so. The expression "commonwealth" or "state of cells" is today no longer to be regarded as a metaphor. The fact remains that:

To declare that of the component cells that go to make us up, each one is an individual self- centered life is no mere phrase. It is not a mere convenience for descriptive purposes. The cell as a component of the body is not only a visibly demarcated unit-life centered on itself. It leads its own life... the cell is a unit life, and our life, which in its turn is a unitary life, consists entirely of the cell lives. (6 — 003)

"Both the pathology of the brain and physiological investigations on sense perception speak unequivocally in favor of a regional separation of the sensorium into domains whose far-reaching independence is amazing because it would let us expect to find these regions associated with independent domains of the mind; but they are not. A particularly characteristic instance is the following. If you look at a distant landscape first in the ordinary way with both eyes open, then with the right eye alone, shutting the left, then the other way round, you find no noticeable difference. The psychic visional space is in all three cases identically the same. Now this might very well be due to the fact that from corresponding nerve ends on the retina the stimulus is transferred to the same center in the brain where "the perception is manufactured" —just as, for example, in my house the knob at the entrance door and the one in my wife's bedroom activate the same bell, situated above the kitchen door. This would be the easiest explanation; but it is wrong. To understand this, consider a thought experiment on the threshold frequency of flickering, a very interesting experiment:

"Think of a miniature lighthouse set up in the laboratory and giving off a great many flashes per second, say forty or sixty or eight or one hundred. As you increase the frequency of flashes, the flickering disappears at a definite frequency, depending on the experimental details, and the onlooker whom we suppose to watch with both eyes in the ordinary way, sees then a continuous light (6-004). Let this threshold frequency be sixty per second in given circumstances. Now in a second experiment, with nothing else changed, a suitable contraption allows only every second flash to reach the right eye, every other flash to reach the left eye, so that every eye receives only thirty flashes per second. If the stimuli were conducted to the same physiological center, this should make no difference: If I press the button before my entrance door, say every two seconds, and my wife does the same in her bedroom, but alternately with me, the kitchen bell will ring every second. However, in the second flicker experiment it is not so. Thirty flashes to the right eye plus alternating thirty flashes to the left are far from sufficient to remove the sensation of flickering; double the frequency is required for that, namely sixty to the right and sixty to the left, if both eyes are open.

"What then is the conclusion?

"It is not spatial conjunction of cerebral mechanism, which combines the two reports...It is much as though the right-eye and left-eye images were seen each by one of two observers and the minds of the two observers were combined to a single mind. It is as though the right eye and left eye are elaborated singly and then psychically combined to one...It is as if each eye had a separate sensorium of considerable dignity proper to itself, in which mental processes based on that eye were developed up to even full perceptual levels. Such would amount physiologically to a visual sub brain. There would be two such sub brains, one for the right eye and one for the left eye. Contemporaneity of action rather than structural union seems to provide their mental collaboration."

A rumble of appreciation from the audience could be heard.

And then this passage was delivered...(6 – 005)

"Are there thus quasi-independent sub brains based on the several modalities of sense? In the roof brain the old "five" senses instead of being merged inextricably in one another and further submerged under mechanism of higher order are still plain to find, each demarcated in its separate sphere. How far is the mind a collection of quasi-independent perceptual minds integrated psychically in large measure by temporal concurrence of experience? When it is a question of "mind," the nervous system does not integrate itself by centralization upon a pontifical cell. Rather, it elaborates a million-fold democracy whose each unit is a cell... the concrete life compounded of sublives reveals, although integrated, its additive nature and declares itself an affair of minute foci of life acting together...when, however, we turn to the mind there is nothing of all this. The single nerve cell is never a miniature brain. The cellular constitution of the body need not be for any hint of it from "mind." A single pontifical brain cell could not assure to the mental reaction a character more unified, and non-atomic then does the roof-brain's multitudinous sheet of cells. Matter and energy seem granular in structure, and so does "life," but not so "mind.""

Another rumble from the audience...gradually increases in magnitude...culminating in a standing ovation...

SS continued: "This too is an arithmetical paradox, a paradox of numbers, and it has, so I believe, very much to do with the one to which I had given this name earlier, though it is by no means identical with it. The previous one was, briefly, the one world crystallizing out of the many minds. And this one is the *one* mind, based ostensibly on the many cell-lives or, in another way, on the manifold sub-brains, each of which has such a considerable dignity proper to itself that we feel impelled to associate a sub-mind with it. Yet we know that a sub-mind is an atrocious monstrosity, just as is a plural mind — neither having any counterpart in anybody's experience, neither being in any way imaginable.

I submit that both paradoxes will be solved by assimilating into our Western build of science the Eastern doctrine of identity. Mind is by its very nature a *singulare tantum*. I should say: the overall number of minds is just one (6-006). I venture to call it

114

indestructible since it has a peculiar timetable, namely, mind is always *now*. There is really no before or after for mind. There is only a now that includes memories and expectations. But I grant that our language is not adequate to express this, and I also grant, should anyone wish to state it that I am now talking religion, not science — a religion however, not opposed to science, but supported by what disinterested scientific research has brought to the fore. It could be that "man's mind is a recent product of our planet's side," but if the first word (man's) is taken out, I would not agree. It would seem queer, not to say ridiculous, to think that the contemplating, conscious mind that alone reflects the becoming of the world should have made its appearance only at some time in the course of this "becoming," should have appeared contingently, associated with a very special biological contraption, which in itself quite obviously discharges the task of facilitating certain forms of life in maintaining themselves, thus favoring their preservation and propagation: forms of life that were latecomers and have been preceded by many others that maintained themselves without that particular contraption (a brain). Only a small fraction of them (if you count by species) have embarked on "getting themselves a brain." And before that happened, should it all have been a performance to empty stalls? Nay, may we call a world that nobody contemplates even that? When an archaeologist reconstructs a city or a culture long bygone, he is interested in human life in the past, in actions, sensations, thoughts, feelings, in joy, and sorrow of humans, displayed there and then. But a world existing for millions of years without any mind being aware of it, contemplating it, is it anything at all? Has it existed? For do not let us forget: to say, as we did that the becoming of the world is reflected in a conscious mind is but a glitch, a phrase, a metaphor that has become familiar to us. The world is given but once. Nothing is reflected. The original and the mirror image are identical. The world extended in space and time is but our representation. Experience does not give us the slightest clue of its being anything besides that.

But the romance of a world that had existed for many millions of years before it, quite contingently, a produced brain, in which to look at itself has an almost tragic continuation that I should like to describe again in the following words:

"The universe of energy is, we are told, running down. It tends fatally toward an equilibrium, which shall be final. An equilibrium in which life cannot exist. Yet, life is being evolved without pause. Our planet in its surround has evolved it and is evolving it. And with it evolves mind. If mind is not an energy-system, how will the running down of the universe affect it? Can it go unscathed? Always so far as we know the finite mind is attached to a running energy-system. When that energy system ceases to run what of the mind, which runs with it? Will the universe, which elaborated and is elaborating the finite mind, then let it perish?

Such considerations are in some way disconcerting. The thing that bewilders us is the curious double role that the conscious mind acquires. On the one hand it is the stage, and the only stage on which this whole world-process takes place. On the other hand, we gather the impression, maybe the deceptive impression that within this world-bustle the conscious mind is tied up with certain very particular organs (brains), which while doubtless the most interesting contraption in animal and plant physiology are yet not unique, not *sui generis*; for like so many others they serve after all only to maintain the lives of their owners, and it is only to this that they owe their having been elaborated in the process of speciation by natural selection.

"Speaking without metaphor, we have to declare that we are here faced with one of these typical antinomies caused by the fact that we have not yet succeeded in elaborating a fairly understandable outlook on the world without retiring our own mind, the producer of the world picture, from it, so that the mind has no place in it. The attempt to press into it, after all, necessarily produces some absurdities.

"Life is valuable in itself. Nature has no reverence toward life. Nature treats life as if it were the most valueless thing in the world. Produced million-fold, it is for the greatest part rapidly annihilated or cast as prey before other life to feed it. This precisely is the master-method of producing ever-new forms of life.

"Thou shalt not torture, thou shalt not inflict pain!" Nature is ignorant of this commandment. Its creatures depend upon racking each other in everlasting strife.

"'There is nothing either good or bad but thinking makes it so.' No natural happening is in itself either good or bad, nor is it in itself either beautiful or ugly. The values are missing, and quite particularly meaning and end are missing. Nature does not act by purposes. Most painful is the absolute silence of all our scientific investigations toward our questions concerning the meaning and scope of the whole display. The more attentively we watch it, the more aimless and foolish it appears to be. The show that is going on obviously acquires a meaning only with regard to the mind that contemplates it. But, what science tells us about this relationship is patently absurd: as if mind had only been produced by that very display that it is now watching and would pass away with it when the sun finally cools down and the earth has been turned into a desert of ice and snow.

"Let me briefly mention the notorious atheism of science, which comes, of course, under the same heading. Science has to suffer this reproach again and again, but unjustly so. No personal god can form part of a world model that has only become accessible at the cost of removing everything personal from it. I do not find God anywhere in space and time—that is what the honest naturalist tells you. For this he incurs blame from him in whose catechism is written: God is spirit.

…Thunderous Applause…

…Tea Break…

Next in line was the revised version of the scientist Freeman Dyson, hereinafter denoted by the symbol "D," who in turn spoke in praise of two great scientists, John H. Wheeler and Eugene Wigner—elaborating their views. In between his speech he also showed a video clip of a conversation between Wheeler and Wigner that took place during the Centennial Symposium, held in 1979, to celebrate the achievements of Albert Einstein.

The presentation carried on until about 1800 hours on day one and was continued the next morning.

An extract of the same is given below:

THE UNIVERSE AS A SELF—EXCITED CIRCUIT

BEYOND THE BLACK HOLE

THE PARTICIPATING OBSERVER

(Extracts from *Some Strangeness in the Proportion* edited by Harry Woolf, including the comments of Freeman Dyson on John Wheeler's presentation on the topic, "Beyond the black hole.")

"About sixty years back, Richard Feynman told me about his "sum over histories" version of quantum mechanics. 'The electron does anything it likes,' he said. 'It just goes in any direction at any speed, forward or backward in time, however it likes, and then you add up the amplitudes and it gives you the wave function.' I said to him, 'You're crazy.' But he wasn't.

"Then about thirty-three years back in 1979 during the Centennial Symposium to Celebrate the Achievements of Albert Einstein, John Wheeler said something similar at an even more basic level. He said, 'The whole universe does what it likes, and then you observe it and it gives you the laws of physics. Freedom from law produces law.' I said again 'You're crazy.' But he wasn't.

"Einstein once wrote of Faraday (translated by Hoffman):

'This man loved mysterious nature as a lover loves his distant beloved. In his day there did not exist the dull specialization that stares with self-conceit through horn-rimmed glasses and destroys poetry.'

"Fortunately for us, dull specialization has never destroyed the poetry of John Wheeler's imagination. Even more than Faraday, Wheeler loves the mysteries of nature and uses the whole universe as his playground. For several years (or rather decades), Wheeler has been astounding the narrow specialists of science with his speculations. The really astounding thing about Wheeler's speculations is that so many of them have turned out in the end to be right. I quote from Wheeler himself his own explanations of how it happens that his imaginative visions often lead to truth: 'We will first understand how simple the universe is when we recognize how strange it is. The simplicity of that strangeness, Everest summit, so well directs the eye that the feet can afford to toil up and down many a wrong mountain valley, certain stage by stage to reach someday that goal.' This is Wheeler's style, a style inseparable from the substance of his thinking, as poetry is inseparable from science in his mind.

"Wheeler's philosophy of science is much more radically relativistic than Einstein's. Wheeler would make all physical law relative to observers. He has us creating physical laws by our existence. In principle, if the role of observers in the universe is as essential as he imagines, life may even create physical laws by conscious decision. This is a radical departure from the objective reality in which Einstein believed so firmly. One of the questions that has always puzzled me is this: Why was Einstein so little interested in black holes? How could he have been so indifferent to the promise of his brightest brainchild? I suspect that the reason may have been that Einstein had some inkling of the road along which John Wheeler was traveling, a road profoundly alien to Einstein's philosophical preconceptions. Black holes make the laws of nature contingent on the mechanical accident of stellar collapse. John Wheeler embraces black holes because they show most sharply the contingent and provisory character of physical law. Perhaps Einstein rejected them for the same reason.

"Let me now come down to the details of what Wheeler has been saying. He says:

'Law without law. It is difficult to see what else than that can be the plan of physics. It is preposterous to think of the laws of physics as installed by a Swiss watchmaker to endure from everlasting to everlasting when we know that the universe began with a Big Bang. The laws must have come into being. Therefore they could not have been always 100 percent accurate. That means that they are derivative, not primary...events beyond law. Events so numerous and so uncoordinated that, flaunting their freedom from formula, they yet fabricate firm form...The universe is a self-excited circuit (6-007). As it expands, cools, and develops, it gives rise to observer-participancy. Observer-participancy — via the mechanism of the delayed-choice experiment (6-008) — in turn gives what we call tangible reality to the universe...Of all the strange features of the universe, none are stranger than these: time is transcended, laws are mutable, and observer participancy matters.'

"The idea of observer-participancy is for Wheeler central to the understanding of nature. Observer-participancy means that the universe must have built into it from the beginning the potentiality for containing observers. Without observers there is no existence. The activity of the observers in the remote future is foreshadowed in the remote past and guides the development of the universe throughout its history. The laws of physics evolve from initial chaos into the rigid structure of quantum mechanics, because observers require a rigid structure for their operations. All this sounds to a contemporary physicist vague and mystical. But we should have learned by now that ideas that appear at first sight to be vague and mystical sometimes turn out to be true.

"Wheeler was building on the work of Bob Dicke and Brandon Carter, who were the first to point out that the laws of physics and cosmology are constrained by the requirement that the universe should provide a home for theoretical physicists. Brandon Carter has shown that the existence of a long-lived star such as the sun, giving steady warmth to allow the slow evolution of life and intelligence, is only possible if the numerical constants of physics have values lying in a restricted range. Carter calls the requirement that the universe be capable of breeding physicists the "anthropic principle." Dicke and Carter used the anthropic principle to set quantitative limits to the structure of the universe. Wheeler carries their ideas much further, conjecturing that the laws of nature are not quantitatively constrained but qualitatively molded by the existence of observers.

"Wheeler unified two streams of thought that had before been separate. On the one hand, in the domain of astronomy and cosmology, the anthropic principle constrains the structure of the universe. On the other hand, in the domain of atomic physics, the laws of quantum mechanics take explicitly into account the fact that atomic systems cannot be described independently of the experimental apparatus by which they are observed. Wheeler has made an interpolation over the enormous gap between the domains of cosmology and atomic physics. He conjectures that the role of observer is crucial to the laws of physics, not only at the two extremes where it has hitherto been noticeable. But, over the whole range, and that the requirement of observability will ultimately be sufficient to determine the laws completely; I think he is probably right, though it may take a little while for particle physicists and astronomers and mathematics to fill in the details in his grand picture of the cosmos.

"That is all I have to say in direct response to Wheeler. The rest of my remarks will be concerned with some speculations of my own that are partly inspired by Wheeler's ideas. I have been concerned as he has with the ultimate role of life and intelligence in the universe. I have not tried as he did to deduce the laws of physics from the existence of life and intelligence. I have only been playing with a much easier problem, trying to deduce from the known laws of physics what the ultimate scope of intelligence and observation may be. As ground rules for my game, I assume that the laws of physics do not change with time and that the relevant laws of physics are already known to us. It is of course highly improbable that the presently known laws of physics are the final and unchanging truth. But we have learned from Wheeler that it is better to be too bold than too timid in extrapolating our meager knowledge from the known into the unknown.

"Looking at the past history of life, we see that it takes about 10^6 years to evolve a new species, 10^7 years to evolve a genus, 10^8 years to evolve a class, 10^9 years to evolve a phylum, and less than 10^{10} years to evolve all the way from the primeval slime to *Homo sapiens*. If life continues in this fashion in the future, it is impossible to set any limit to the variety of physical forms that life may assume. What changes could occur in the next 10^{10} years to rival the changes of the past? It is conceivable that in another 10^{10} years life could evolve away from flesh and blood and become embodied in the interstellar black cloud, as described by Fred Hoyle. Or in a sentient computer, as described by Karel Capek.

"Is the basis of consciousness matter or structure? This is a deep question that we do not know how to answer. Let me spell out its meaning more explicitly. My consciousness is somehow associated with a collection of organic molecules inside my head. The question is whether the existence of my consciousness depends on the actual substance of a particular set of molecules, or whether it only depends on the structure of the molecules. In other words, if I could make a copy of my brain with the same structure but using different materials, would the copy think it was me? I assume as a working hypothesis that the answer to this question is yes that the basis of consciousness is structure. Then it is possible to talk about life and intelligence in abstract terms, independent of the details of organic chemistry and the physiological properties of flesh and blood.

"How can we characterize quantitatively the structural essence of a living creature independently of its substance? One attribute of a creature that may be useful for this purpose is a number Q that I call the "complexity" of the creature. Q is the quantity of entropy that the creature produces through its metabolism during each instance of consciousness. If entropy is measured in information units, or bits, Q is a pure number without physical dimensions, expressing the amount of information that must be processed in order to keep the creature alive long enough to say "Cogito, ergo sum." For example, a human being dissipates about two hundred watts of power at a temperature of three hundred degrees Kelvin, with each moment of consciousness lasting about 1 second. A human being therefore has complexity Q of the order of 10^{23}. It is probably no accident that this number is of the same order of magnitude as the number of molecules in the human brain. The hypothesis that the basis of consciousness is structure rather than substance can then be stated quantitatively as follows. If A and B are two material embodiments of life, and a creature exists in A with a certain complexity Q, then a subjectively equivalent creature can exist in B with the same Q.

"This system of life's adaptability may sound pale and abstract, lacking in color and detail. But it has far reaching consequences. It implies that the universe becomes more and more hospitable to life as space becomes colder and quieter. It implies that the prime requirement for life is not an abundance of energy but a good signal-to-noise ratio. It means that, at least if we are lucky enough to be living in an open universe that continues to expand forever, the material basis exists for life also to persist and grow indefinitely. A society of any given complexity can survive forever and constantly increase its store of information and experience, using a finite total quantity of energy. The actual amount of energy required for life to exist forever turns out to be less than 0.01 joule per unit of complexity.

"In conclusion, I would like to put Wheeler's speculations about the role of observers in determining the laws of physics together with my own speculations about the inexhaustible potentialities of life in the universe. According to Wheeler, the laws of physics evolve progressively in such a way as to make the universe observable. According to me, the scope of the observer expands forever as time proceeds. These two speculations together give support to the conjecture that the laws of physics themselves are inexhaustible. Fortunately, the ideas of Wheeler give us a basis for believing that the world of physics may indeed be truly inexhaustible. Especially if the Universe is open and infinite in time, the world of life and consciousness is inexhaustible too. Then there will always be new worlds to explore, new effects of observer-participation reacting back upon the laws of physics, new connections for the Einstein's and John Wheelers of the future to speculate upon.

**

There was a break in Freeman Dyson's talk. In fact, the beginning of this break coincided with the end of the first day's proceedings.

During this break—the first half hour of the second day—Dyson showed a video clip of John Wheeler and Eugene Wigner engaged in an extremely interesting conversation on the highly complex topic of black holes and their information contents, the discussion leading up to the delayed choice experiment and the strange things that quantum mechanics does to causation. (This conversation actually took place during the 1979 symposium referred to earlier.) (Refer 6 -009)

Dyson then comes back to talk about Eugene Wigner and his philosophy that the consciousness of the mind itself is the hidden variable that explains the collapse of wave (phenomena).

In Wigner's own words:

"...the result of an observation modifies the wave function of a system. The modified wave function is, furthermore, in general, unpredictable before the impression gained at the interaction has entered our consciousness: it is the entering of an impression into our consciousness, which alters the wave function because it modifies our appraisal of the possibilities of different impressions, which we expect to receive in the future. It is at this point that the consciousness enters the theory unavoidably and unalterably."

The talk also includes the famous "Wigner's friend" in relation to the "Schrödinger's cat thought experiment" (described in the next speaker's presentation).

The next speaker (on day two of the seminar) was John Cramer—a scientist who conceptualized a unique interpretation of quantum mechanics, based on the ideas of Feynman and Wheeler (Wheeler-Feynman absorber theory) regarding advanced waves that go into the past and retarded waves that go into the future. The topic for his talk was: "Quantum Particles Shake Hands in Agreement."

Quantum entanglement once again took center stage in this talk, which included Cramer's "transactional interpretation of quantum mechanics," an interpretation that explains entanglement, but does not require the violation of Einstein's special theory of relativity. Later- on day three of the seminar—the linkage with "Six Words" would be shown, and how nicely the "Six Words" together with Wigner's "consciousness philosophy" fit into the entire picture.

This section is enormously complex. And, this is an understatement.

If I were to describe in full details what transpired during this session wherein the speaker described his "controversial" interpretation of quantum mechanics and the hue and cry that followed in the Q and A session, I will be taking a major risk in that the reader may put down this autobiography and never pick it up again. However as in the case of chapter four on the subject of "Bell's inequality," I will endeavor to incorporate this in the second edition of the book. For the present, I will just give a brief account of what this interpretation is all about. In case the reader decides to skip even this, I will not blame him for this act. For those readers who would like a more comprehensive coverage, I recommend they should read John G. Cramers paper on the subject, published in July 1986 in *Reviews of Modern Physics*.

To begin, let me give a summary of another interpretation of quantum mechanics *viz* "the guide wave interpretation of de Broglie":

"...a particle is a very small object, which is constantly localized in space and a wave is a physical process, which is propagated in space in the course of time, according to a given equation of propogation...The wave has a very low amplitude and does not carry energy, at least not in a noticeable manner. The particle is a very small zone of highly concentrated energy incorporated in the wave, in which it constitutes a sort of generally mobile singularity. By reason of this incorporation of the particle in the wave, the particle possesses an internal vibration, which, as it moves, remains constantly in phase with the vibration of the wave...the mean path of the particle is determined according to the shape of the wave by a certain "guidance law," but this motion has superimposed on it continual fluctuations corresponding to a hidden variable behavior of the particles."

De Broglie's interpretation is said to be responsible for stimulating the development of quantum physics (via the Schrödinger equation) as well as for the emphasis on the wave aspects of quantum mechanics. While the Guide Wave Interpretation of de Broglie has certain shortcomings such as the fact that it makes no provisions for nonlocalities and is inconsistent with the Bell inequality experiments, it can be considered a precursor to the transactional interpretation given below:

Transactional Interpretation of Quantum Mechanics

(Extract from "Q is for Quantum" by John Gribbin)

"When an electron vibrates, it attempts to radiate by producing a field, which is a time symmetric mixture of a retarded wave propagating into the future and an advanced wave propagating into the past. As a first step in getting a picture of what happens, ignore the advanced wave and follow the story of the retarded wave. This heads off into the future until it encounters an electron, which can absorb the energy being carried by the field. The process of absorption involves making the electron that is doing the absorbing vibrate, and this vibration produces a new retarded field, which exactly cancels out the first retarded field. So in the future of the absorber, the net effect is that there is no retarded field.

"But the absorber also produces a negative-energy advanced wave traveling backward in time to the emitter, down the track of the original retarded wave. At the emitter, this advanced wave is absorbed, making the original electron recoil in such a way that it radiates a second advanced wave back into the past. This new advanced wave exactly cancels out the "original" advanced wave, so that there is no effective radiation going back in the past before the moment when the original emission occurred. All that is left is a double wave linking the emitter and the absorber, made up of half of a retarded wave carrying positive energy into the future and half of an advanced wave carrying negative energy into the past (in the direction of negative time), in line with the "Wheeler-Feynman absorber theory."

"Because two negatives make a positive, this advanced wave adds to the original retarded wave as if it too were a retarded wave traveling from the emitter to the absorber. In Cramer's words: 'The emitter can be considered to produce an 'offer' wave, which travels to the absorber. The absorber then returns a 'confirmation' wave to the emitter and the transaction is completed with a 'handshake' across space-time. But this is only the sequence of events from the point of view of pseudo time. In reality the process is atemporal; it happens all at once.'" (ref 6-009a)

"In order to illustrate this interpretation, let us study it in the light of the other interpretations of quantum mechanics with special reference to their shortcomings and judge for ourselves if these can be taken care of by the transactional interpretation. For this, let us go back to the Schrödinger's Cat thought experiment and study that in relation to another thought experiment devised by Wigner called "Wigner's friend."

(Extract from John G. Cramer's paper on "The Transactional Interpretation of Quantum Mechanics):

"The experimenter places the cat inside the box, seals it, and leaves the system undisturbed for an hour. At the end of the hour the experimenter deactivates the counter, opens the box, and observes the state of the system. Two states are possible: a state AC — a live cat in which the flask is unbroken and the cat remains alive, and a state DC — dead cat in which the flask has shattered and the cat has been killed. Schrödinger's question is: What is the quantum mechanical state vector of the system immediately before the box is opened and the observation is made? As per the Copenhagen interpretation the SV of the system consists of equal components of the live-cat wave function and the dead-cat wave function, until such time as the observer collapses the SV into one or the other of these states by making an observation, since it is the change in the observer's knowledge, which precipitates the SV collapse. In the period just before the observation is made, the SV describes the cat as 50 percent alive and 50 percent dead. This description, which may seem plausible enough when applied to a microscopic system, appears rather absurd when applied to an individual complex organism like a cat. Wigner further heightened the weirdness implicit in the Copenhagen interpretation by replacing the cat with a "friend," an intelligent observer, and at the same time replacing the prussic acid mechanism with a less lethal piece of apparatus, a light bulb, which is switched on when a count is recorded. The experimenter then performs the experiment, which can be considered as two experiments: 1) treating the friend + box as a system, the experimenter makes an observation, and 2) treating the counter mechanism as a system, the friend makes an observation, which is subsequently reported to the experimenter.

Wigner's conclusion: "Consciousness must have a special role in the collapse of the SV, for otherwise one must deal (at least on the philosophical level) with uncollapsed SVs containing conscious observers in a multiplicity of alternate states. Several others have suggested alternative ways of avoiding uncollapsed SVs, describing conscious observers. Heisenberg suggested that the SV collapses when the system enters the domain of thermodynamic irreversibility, e.g., as soon as a piece of macroscopic apparatus becomes involved. Schrödinger suggested that as soon as a permanent record of the system state is made, e.g., by smashing the flask, the SV is collapsed. Hugh Everret dispatched the interpretational problem posed by these thought experiments by suggesting that the SV never collapses. Instead the universe "splits" with each quantum event into alternate universes, each characterized by one of the possible outcomes of the event.

"The central focus of the problems posed by Schrödinger's cat and Wigner's friend is the question of when the SV actually collapses. Is it when the gamma ray leaves the radioactive nucleus? Is it when it interacts with the Geiger counter? When the flask is smashed? When the cat dies? When the observer looks in the box? When he tells a colleague what he observed? When he publishes his observations in the *Physical Review*? When? A billiard shot is over when the billiard balls stop colliding and come to rest. But the atomic "billiard balls" of a quantum billiard shot continue to collide forever, never coming to rest so that the shot can be considered finished. The transactional interpretation avoids the implicit dilemma because in the TI the SV collapse, i.e., the formation of the transaction, is atemporal. During the entire one-hour period that the box is closed, the radioactive source S of Schrödinger's apparatus sends out a very weak offer wave (OW). This OW may or may not, with equal 50 percent probabilities, be confirmed by a confirmation wave from the Geiger counter so that a completed transaction is formed. If a transaction is formed, then the count is recorded, the flask shattered, the cat killed. If such a transaction is not formed then the cat remains alive. The SV (or OW) does indeed have implicit in it both live-cat and dead-cat possibilities, but the completed transaction allows only one of these possibilities to become real. Because the collapse does not have to await the arrival of the observer, there is never a time when "the cat is 50 percent alive and 50 percent dead." And the need for consciousness, permanent records, thermodynamics, or alternate universes never arises. The "buck stops" at the absorber, in this case the Geiger counter, and the uncollapsed SV need not be tracked any further."

This talk took up the remainder of the pre-lunch session of Day two.

**

For the post-lunch session of day two, three speakers, who were in fact the revised versions of the three great philosophers Rene Descartes, Immanuel Kant, and Arthur Schopenhauer took the stage one after the other and spoke on the subject:

THE PHILOSOPHY OF SCIENCE

"Whatever their differences of opinion about the nature of God, I know of no religion that does not teach that God is a mind."
God and the New Physics by Paul Davies
"The human mind inevitably imposes order on the world so as to make sense of it." Immanuel Kant
No scientist of any substance could have formulated their philosophies without an in-depth study of the writings of these three great philosophers referred above. It has been said by someone — who read half of Immanuel Kant's *Critique of Pure Reason*" (a path-breaking work on philosophy) — while talking to his friends: "One who has not read the book can consider himself to be just a child." When questioned why he did not complete the book, he replied: "It is possible I might have become insane if I had done so."

Now these three speakers discussed and debated the burning question of what is mind and what is soul. Given below is a small extract from what Rene Descartes (revised version, of course) spoke: (6-010)

I THINK THEREFORE I AM

"The human being consists of two distinct, separate kinds of things: the body and the soul, or mind. The body acts as a sort of host or receptacle for the mind, or perhaps even as a prison from which liberation may be sought through spiritual advancement or death. The mind is coupled to the body through the brain, which it uses (via the bodily senses) to acquire and store information about the world. It also uses the brain as a means to exercise its volitions, by interacting on the world. An important feature of this picture is that the mind is a thing; perhaps even more specifically, a substance. Not a physical substance, but a tenuous, elusive, ethereal sort of substance, the stuff that thoughts and dreams are made of, free and independent of ordinary ponderous matter.

The essential notions seem to be: first that there are two distinct orders of being or substances, the mental and the material. Mind or mental substance is neither perceptible by the senses nor extended in space; it is intelligent and purposive and its essential characteristic is thought or rather consciousness. Though the human body is an engine, it is not quite an ordinary engine, since some of its workings are governed by another engine inside it — this interior governor — engine being one of a very special sort. It is invisible, it is inaudible, and it cannot be taken to bits and the laws it obeys are not those known to ordinary engineers."

Question from the audience: "Where exactly is the engine located in the body?"

RD: It could be that this governor is a "ghost in the machine." The soul's insubstantial quality would appear to be necessary for two reasons. First, we do not see souls or detect their physical presence in any direct way, nor are they revealed during brain surgery. Secondly, the world of matter must comply with the laws of physics, which on the macroscopic level, ignoring quantum effects (6-011), are deterministic and mechanical and hence incompatible with free will.

"However, as you all know, free will is a fundamental attribute of the soul. This leads to a substantial paradox; if it is matter at a macroscopic level it cannot have free will, and if it is not matter, then how can it be an engine that is supposed to act as the prime mover of the component parts of the body? In my opinion, this paradox can be resolved by considering this engine to be a quantum-level matter, too infinitesimal to be visible or audible but significant enough to be completely indivisible and to follow such laws, which are not known to ordinary engineers. Perhaps it is a particle—with these sublime properties—manufactured and evolved during its journey through the cosmos in the aftermath of supernovas. All it needs then is a certain biochemistry, to become life, and on its way it picks up a part of this thing called *mind*, which is, of course, ever-present throughout space and time.(6-012).

"Which means that this engine called soul—which is distinct from mind—is a thing in itself; it is at a quantum level, it is invisible, inaudible, but indivisible and has unique properties unlike any other kind of matter whether macroscopic or quantum. With all this, it becomes imperative that it must have a location both in time and space. In my opinion, it is located in the small pineal gland in the brain and provides the elusive physical link between mind and brain. That location is as far as "space" is concerned; as regards "time," it could be that it remains inside the body as long as the biochemistry inside the body is conducive for it to be conscious of itself with the help of the mind. When it is no longer the case, it comes out of the body (when the individual dies), and then looks for another appropriate biochemistry at another time and another place. The mind, however, remains the same except that at any stage the consciousness always remains in the singular.

**

The three philosophers enthralled the audience during this post-lunch session of the second day with their spellbinding talks...

Immanuel Kant (R– 01) spoke for nearly an hour or so. As I mentioned earlier, not many can understand his language. It's stupendously complex. But, in this august gathering of the great scientists and philosophers, including the highly knowledgable delegates, there were quite a few who could understand him (yours truly was not one of them)…and to say that they were spellbound would be an understatement. I cannot risk putting the *summmom bonnum* of his talk over here for obvious reasons… but I would like to include a small extract from the parts, which I could understand somewhat. The subject was: "Of the Concept of an Object of Pure Practical Reason."

"Now, since the notions of good and evil, as consequences of the *a priori* determination of the will, imply also a pure practical principle, and therefore a causality of pure reason; hence they do not originally refer to objects (so as to be, for instance, special modes of the synthetic unity of manifold of given intuitions of one consciousness) like the pure concepts of the understanding or categories of reason in its theoretic employment; on the contrary, they presuppose that objects are given; but they are all modes (modi) of a single category, namely that of causality, the determining principle of which consists in the rational conception of a law, which, as a law of freedom, reason gives to itself, thereby *a priori* proving itself practical. However, as the actions on the one side come under a law, which is not a physical law, but a law of freedom, and consequently belong to the conduct of beings in the world of intelligence, yet on the other side as events in the world of sense they belong to phenomena; hence the determinations of a practical reason are only possible in reference to the latter, and, therefore in accordance with the categories of the understanding; not indeed with a view to any theoretic employment of it, i.e, so as to bring the manifold of (sensible) intuition under one consciousness a priori…But only to subject the manifold of desires to the unity of consciousness of a practical reason, giving it commands in the moral law, i.e., to a pure will *a priori*."

I could make some sense from the above…but I am not sure if I am right…so I am putting it in a footnote, which also includes an interaction with a blogger (and a friend) on the subject "difficult books." (Refer 6-014)

Then came Amaury de Riencourt, and, after showering prais-
es on the three philosophers, he spoke at length on "Eastern mysticism and science":

SQUARING THE CIRCLE—BEYOND THE MIND

(Extracts from *The Eye of the Siva*" by Amaury de Riencourt)

"The Western intellect has come to the conclusion—as nicely
explained by Kant—that ultimate reality is simply ungraspable.
The intellect can deal with the world of the phenomenon but
not with the reality underlying it, the mysterious background
or substratum of the *neumenon*, the "thing-in-itself." Kant has
asserted that a part of knowledge is extra-intellectual, but has
refused to deal with this *a priori,* claiming that the framework
of understanding must be accepted as it is, readymade. Kant,
however, has overlooked the fact that logical thought and cre-
ative experience move in two contrary directions—one, according to the rational intellect, which, geared to act on the objective
world of matter and partaking of its structure, descends toward
it and turns its back on life, the other moving in the same direction as life, involves the whole of consciousness with its
intuitive component. Kant thinks of intuition as being *infra*-in-
tellectual—whereas I think it is *supra*-intellectual, overflowing
and surrounding on all sides the narrow limits of the rational
mind as the iris of the eye surrounds the pupil. This is the fundamental discovery of Eastern thought. The sense perceptions
are superior to the physical body, the mind is superior to the
senses, intuitive understanding is superior to the mind, and
above all the self is superior to intuitive understanding. (Refer
to the Bhagavad Gita.)

"It is interesting to note that neurophysiology now tells us
that the two hemispheres of the brain each specialize in different
functions of cognition—the left dedicating itself to logical and
analytical thinking, seat of verbal thought; the right to intuition
and holistic understanding of patterns, with an ability to grasp
directly the relationship between parts of the whole. One operates linguistically with rational sequences of deductions and

inductions; the other intuitively juxtaposes images and symbols, integrates and synthesizes rather than analysis. It seems obvious that Western philosophic and scientific culture has, so to speak, atrophied the right hemisphere of the brain, while Eastern culture has given it predominance — or rather it might well be the other way around in the sense that the cultural divergence between East and West would have its origin in the physiological differentiation in the brain. The proper role of metaphysics is therefore to use the whole of consciousness with its intuitive component in order to insert itself in the living evolutionary movement of becoming instead of recreating it artificially with the dead fragments of itself. We must strive to achieve a coincidence (transformation) of individual consciousness with the living principle that generates it."

**

"This brings us to the threshold of the much-discussed problem of freedom. What is the true nature of "I?" On the surface that is in the conscious mind geared to act on the external world, and under the impact of sense perceptions there is indeed associationism of juxtaposed terms. But in depth, in consciousness and in the unconscious, there is less and less juxtaposition. States of consciousness interpenetrate one another; and the deeper we go, the more the whole personality is involved in decision making... The determinist illusion springs from an arbitrary effort of abstraction, from an artificial identification of "I" that thinks only with rational intellect, with the one that feels and acts from the depth of his being...The opponents of free will believe that choice is an oscillation in space, whereas it is really a progression in time in which the "I" and its consciousness are in a perpetual becoming. The "I" feels its freedom and acts upon it, but cannot explain it rationally because it has to refract it through a spatial scientific determinism. Freedom is not analyzable, being the relationship between the concrete "I" and the act that is being accomplished. One can analyze a thing but not a progression, one can cut up an extension but not duration...So it is that just as the philosophic contradictions such as freedom versus determinism, for instance, cannot be resolved logically, but can, in fact, be overcome experimentally, so can this quest for ultimate knowledge, and therefore realization, of the self, be successful by overcoming the rational mind, thus disconnecting the subject from the phenomenal world of time and space.

In actual fact, this squares the circle that can never be squared by strictly intellectual means."

**

In the said parallel universe, then came the scientist Paul Davies considered by many as one of the most adept science writers on either side of the Atlantic, with his awe-inspiring talk on science and religion, and his elaborate explanation of his contention that the new physics offers a much surer path to God than religion. In the end he made the following statement, of which the authenticity and impact was unquestioned: "Whatever their differences of opinion about the nature of God, I know of no religion that does not teach that God is a mind."

The great scientists and the great philosophers had made their points. In general they were complimentary. With some contradictions…

But this universe is still in its infancy (in accordance with our time) and there is ample time to resolve the contradictions and arrive at the final truth. Until such time we need to live with all the contradictions and still arrive at the principles that we must follow in order to survive the next three hundred years considered as "the danger zone."

And thus, the first two days of the seminar were concluded.

**

Sleepless night ensued…I needed that someone or something to reenter my head and help me on the morrow…

Lightning struck…when it happened…and then enlightening struck…(Refer 6-015)

ENDNOTES (6-001 TO 6-015)

6-001:

Anthropomorphism: The attribution of human characteristics to something that is not human, for instance, to the weather, or to God.

6-002:

Antinomies: Contradictory conclusions from equally good premises.

6-003:

The portion of the extract written in bold italics correspond to passages quoted by Schrödinger from Charles Sherrington's *Man on His Nature*.

6-004:

In this way the fusion of successive pictures is produced in the cinema.

6-005:

Passage par excellence. Read this again and again. A unique paragraph whose interaction on me was utterly spellbinding.

6-006:

Singulare Tantum – This corresponds to the ten words of Schrödinger, which are in conformity with the equation of the Upanishads, and have the same meaning as the "six words." See also chapter Six E.

6-007 and 6-008:

The universe As a Self-Excited Circuit:

Starting small, it grows (loop of U) and in time gives rise to observer-participancy, which in turn imparts "tangible reality" to even the earliest days of the universe.

In order to understand this we must first understand the mechanism of an experiment called the "delayed choice experiment." This is a thought experiment conceptualized by John Wheeler. To grasp this, let us go back to the double slit experiment with electrons. Consider that a new detector screen is set up somewhere in between the two holes and the last detector screen. Now in accordance with the quantum theory, if we decide to turn off this new detector there will be an interference pattern formed in the main detector. On the other hand if we decide to *delay* our choice of looking, and look only after the electron has passed the two-hole screen by activating/or installing the new detector — by means of a fast computer — only at that stage do we find that there is no interference pattern. Actual experiments have in fact been carried out at the Universities of Maryland in the United States and Munich in Germany to confirm this astounding phenomenon. How does this happen? We ask again what in the world is going on and what does this mean? This delayed decision seems to have affected how the electron behaved when it was in the process of passing through the hole(s) a tiny fraction of a second (perhaps a billionth of a second) in the past.

Now, this idea of the universe as a self-excited circuit was conceptualized by Wheeler on the basis that the timescales involved need not be so tiny, and we can imagine a similar experiment on a literally cosmic scale. Consider, for example, light from a distant object such as a quasar, which arrives after having traversed two different routes after getting bent around a massive galaxy in the line of sight — the phenomena known as gravitational lensing — the two quasar images can be combined to make an interference pattern to prove that the light traveled across the universe like a wave following both routes. However, if we monitor individual photons to check which route they traversed, there is no such interference pattern. Now the quasar might be ten billion light-years away, but we have timed our observation today and it has affected the way the light set out on its journey ten billion years ago.

In the words of John Wheeler:

"There is nothing in principle to prevent it (the time between the action and our observation of the action) being a billion light-years in the sense that we in the here and the now, by observing the photon of the primordial cosmic fireball radiation, have an irretrievable consequence for what we have the right to say about that photon. In that sense we are operating in a reverse direction of the sense of time and it is in that sense that anything having to do with things swimming up over the horizon also comes into this category of events. And it is in this sense that quantum mechanics does strange things to what we call causality if we examine it with sufficient care."

For further reading:
Ideas and opinions by Albert Einstein
What Is Life? by Erwin Schrödinger
Man on His Nature by Charles Sherrington
Quantum Questions by Ken Wilber
Some Strangeness in the Proportion edited by Harry Woolf.
Q is for Quantum by John Gribbin

6-009:

The Conversation:

"Wigner: I have a question within physics that always bothers me in connection with the black hole, and it is this. The laws of relativity are time inversion invariant. As a result, it is clear that the same amount of information is contained after the black hole collapses as before. This is not evident because just as, even though the laws of physics are also time inversion invariant, thermodynamics shows that things come to equilibrium. But in that case we can see what it means. It means that the information that time reversal will push the system back into its original state is not relevant because we cannot reverse the velocity of every particle. Now the question is: What is a similar problem in the black hole situation? What does become uninteresting? Tell us, John.

Wheeler: One has to recognize right from the start that in all the theory of black hole physics that we do today, we feed in the assumption about the direction of time—we do not deduce it. We put it in as an initial condition. We could perfectly well talk of holes that work the other way round, just as we could perfectly well talk of heat going from a cold teacup to a hot teacup. But it would violate all our understanding that it is initial conditions that count and not final conditions. And so, this one-sidedness we feed in as a generalization of our experience—not understanding, at least I don't understand—why the world is built that way, with a one-sided initial condition.

Wigner: In thermodynamics we say, when deriving the law of entropy that the initial conditions are irregular except for those, which are macroscopically constrained. Otherwise initial conditions are fully irregular. This is the basis of the increase of entropy. What is a similar postulate for the black hole? Clearly since the laws are reversible, the final information is just as great as the initial one.

Wheeler: It is that the final information has become useless: it's. . .

Wigner: But in what way? I mean, what part of it is useless?

Wheeler: It has gone behind the horizon where we cannot grab hold of it.

Wigner: I did not see a horizon.

Wheeler: That is to say, when everything falls in, we cannot get any signal back out again. So we have no possibility of looking at the state of the system. We cannot look at the details of what has gone inside of the horizon of the black hole: all the matter that has gone in, all the particularities it has carried with it, all the complications that are there. Or, if we want to put it, not in terms of the frame of reference of the material falling in, which goes beyond the horizon, but if we want to put it in terms of the standpoint of the person who is looking at it all from outside, then we recognize that everything approaches asymptotically to the horizon of the black hole, to the surface of the black hole, and all these details get washed out.

Wigner: What do you mean by details? That is the question.

Wheeler: The churning about of the molecules, the atoms, the radiation, everything that has fallen in, all the complications of the star, or, if it is a galaxy...

Wigner: Complications—that is not contained in the equation. The equation has gik in it. What happens to the gik that makes it so uninteresting?

Wheeler: Yes. Well, all the occupation numbers of the various quantum states.

Wigner: There are no quantum states in general relativity.

Wheeler: Yes. Well, we are talking about the states of motion of these particles that are going in.

Wigner: I know the theory of—perhaps I should stop—but...

Wheeler: I know this famous statement—Never listen to anything Eugene Wigner says, but when he asks a question, listen very carefully.

Wigner: This makes it easy for me not to speak tomorrow.

Dirac: This discussion had better be completed in private. (Everyone in the audience says, "No!") Well then, let it go on.

Wigner: You see, I understand the increase of the entropy and the approach to equilibrium in a gas that has originally different temperatures at different locations, and one can specify what is becoming uninteresting. Namely, what becomes uninteresting is the structure of an increasing number of domains of phase space. Originally, if I take a small part of phase space, it will be either occupied or unoccupied., but the occupied regions will spread out over the whole space and if I want to see an unoccupied region, I have to go very, very carefully to tiny fractions of phase space. What is the analog of this in the collapse?

Wheeler: Everything that you have dropped in of all the matter that has fallen in is there around the horizon if you will, and . . .

Wigner: What is the horizon?

Wheeler: That is the point $r = 2M$

Wigner: Oh! Oh, good.

Wheeler: The Schwarzschild radius…Everything is coming up to the Schwarzschild radius and decreasing its distance from the Schwarzschild radius by a factor e in every 10^{-5} seconds, so it is an unbelievably compact amount of material there as viewed from a faraway observer, and the possibility of his seeing or ever hoping to know all those details has got lost, so that is where the information has disappeared.

Wigner: Is it—I'll ask one more question, then I'll shut up— is it possible to imagine the explosion of a black hole? I often think that the Big Bang was an explosion of a black hole. Is that nonsense?

Wheeler: Zeldovich and Novikov proposed it some years ago— the idea that there are in addition to black holes, white holes. And that they should vomit forth matter the way a black hole sucks in matter. But then, in the meantime, Eardley has shown that such things are so unstable that in the very earliest 10^{-5} seconds of the universe, such a thing would have blown up and nothing of that kind would remain until today…In principle, if one were to collide two black holes sufficiently soon after the act of formation—within 10^{-5} seconds then there should be enough detail left. However, all the analysis that is conducted today, for example, by Smarr on the collision of two black holes, presupposes a time lapse great enough to have washed away all these details because, after all, one day is something like 10^{10} powers of e, and if you have washed away a detail of 10^{10} powers of e, there is not much left of it to show up in the collision.

NE"EMAN: To the outside observer it is also washed away?

Wheeler: Yes. It is to the outside observer. It has disappeared.

Ostriker: A question to John Wheeler on causality. It seems to me that the universe is open, on the best astronomical evidence. Thus things are always appearing over the horizon that simply by causality we never could have affected. There has not been time for a signal to get from us to them and for the results to get back to us. So that it seems very difficult, if the universe is open, for participatory democracy not to violate causality in the usual way. Is that true or false?

Wheeler: Far more in violation of any ordinary notion of causality and yet again in the end not in violation of it, is this delayed choice double slit experiment, where a choice made in the here and the now produces an irretrievable effect on the past as to whether the electron or photon went through both holes in the metal plate or through only one. The electron or photon has already gone through the metal plate at the moment when you decide, and yet despite that fact, you can have your choice as to which way you will have it. It is in this sense that quantum mechanics does strange things to what we call causality if we examine it with sufficient care.

Ostriker: I'm afraid I do not follow the relevance of that example to the question with regard to galaxies appearing at the horizon — how it would be possible for an observer to influence them?

Wheeler: I have mentioned the delayed-choice double slit experiment but perhaps I have not spoken sufficiently vividly about it in the sense that the experiment in which we were talking about it was the one in which the interval between the metal plate with two holes in it and the place with the venetian blinds where we make our choice — we'll say one foot or time of one nanosecond — but there is nothing in principle to prevent it being a billion light-years in the sense that we in the here and now, by observing a photon of the primordial cosmic fireball radiation, have an irretrievable consequence for what we have the right to say about the photon. In that sense, we are operating in the reverse direction of the sense of time and it is in that sense that anything having to do with things swimming up over the horizon also comes into this category of events.

6-009a:

"If there is one particular link in the event chain that is special (says Cramer), it is not the one that ends the chain. It is the link at the beginning of the chain when the emitter, having received various confirmation waves from its offer wave, reinforces one of them in such a way that it brings that particular confirmation wave into reality as a completed transaction. The atemporal transaction does not have a 'when' at the end. This dramatic success in resolving all of the puzzles of quantum physics has been achieved at the cost of accepting just one idea that seems to run counter to common sense — the idea that part of the quantum wave really can travel backward through time. At first sight, this is in stark disagreement with the commonsense intuition that causes must always precede the events that they cause. But on closer inspection it turns out that the kind of time travel required by the transactional interpretation does not violate the everyday notion of causality after all—and nor does all of this atemporal handshaking across the universe necessarily remove that most prized of our human attributes, our freedom of will.

6-010:

These are some extracts from the chapter on "Mind and Soul" from the book *God and the New Physics* by Paul Davies who in turn referred to the ideas of Gilbert Ryle and R.J. Hirst.

6-011:

The revised version of RD is up to date with his knowledge of quantum physics and cosmology and like all others in the dais as well as in the audience has read the first five chapters of "Six Words."

6-012:

Refer to chapter two on "My resume—the past—from time zero to time "now"—the package of speculations.

6-013:

That line..."There is someone in my head but it's not me" is needed to explain how I, an ordinary laymen with an average sort of quality and configuration of the brain cells supplied to me by the God who plays dice, is here talking about such complex and serious issues such as saving our human civilization from self-destruction.

And this idea of "something or someone in my head" was generated during a satirical response to a blog by a scientist, Seth Shostak, entitled "The UFO Bestiary" posted on 04/27/2012 on Huffingtonpost.com where "yours truly" claimed to be someone offloaded from a UFO.

A small extract from the response:

"Give me Seth Shostak...your email id...so that I can let you know that...not only did I see a UFO...I was in fact in it.

"I am from Planet X of Star Y, but those are names given by us. I have been forbidden to reveal the "Earth names" by my bosses. I am currently observing what's going on here...And I tell you it's not good. Unless there is a sea change in attitudes...Conflicts... Wars...Population explosion...I give you just five hundred years before extinction time.

6-014:

I recall a conversation with a blogger "Dave Astor" in response to his blog "Some 'Difficult' Books Aren't as Difficult as We Think"...Posted: 02/07/2013 2:39 p.m.

To read the blog refer to www.huffingtonpost.com/ DaveAstor/Some difficult books aren't as difficult as we think.

My comment:

I don't know about fiction, but in nonfiction (other then technical books in specialized subjects), the most difficult book I have come across so far is *Critique of Pure Reason* by Immanuel Kant.

I have little doubt that most...if not all...of those who are in possession of this book will agree with me.

It is possible that no one has read the book in completion...or maybe it is a single-digit number.

It is learned that a person who read about half the book... and understood it well...presumably after ten or more readings...once said: "Whosoever has not read the book is still a child." When asked why he did not complete the book...he promptly replied: "I would have surely become insane if I had done so."

I have this book and I can reproduce some typical paragraphs in support of what I have stated...if Dave wants me to.

Dave Astor in reply to: SKSagar:

Nice to hear from you, SK!

Thanks for mentioning an example of difficult nonfiction! That would also make a great article in itself.

Fascinating comment, with wry moments. If you want to post a paragraph that illustrates the difficulty of the Kant book, please do! I'd love to.

SKSagar in reply to: Dave Astor:

Thanks…Dave

Here is one such paragraph…a simpler one:

"Now, since the notions of good and evil, as consequences of the *a priori* determination of the will, imply also a pure practical principle, and therefore a causality of pure reason; hence they do not originally refer to objects (so as to be, for instance, special modes of the synthetic unity of manifold of given intuitions of one consciousness) like the pure concepts of the understanding or categories of reason in its theoretic employment; on the contrary they presuppose that objects are given; but they are all modes (modi) of a single category, namely that of causality, the determining principle of which consists in the rational conception of a law that as a law of freedom, reason gives to itself, thereby a priori proving itself practical. However, as the actions on the one side come under a law, which is not a physical law, but a law of freedom, and consequently belong to the conduct of beings in the world of intelligence, yet on the other side as events in the world of sense they belong to phenomena; hence the determinations of a practical reason are only possible in reference to the latter, and, therefore in accordance with the categories of the understanding; not indeed with a view to any theoretic employment of it, i.e, so as to bring the manifold of (sensible) intuition under one consciousness *a priori*…But only to subject the manifold of desires to the unity of consciousness of a practical reason, giving it commands in the moral law, i.e., to a pure will *a priori*."

The subject was: "Of the Concept of an Object of Pure Practical Reason."

The paragraph occupied about one-third of a page of the book, and there are 613 pages.

Dave Astor in reply to: SKSagar:

And that's one of Kant's simpler paragraphs? Wow! Thanks, SK.

SKSagar in continuation:

How did I get the book?:

It was a Sunday...Last October...We (with wife and son) were strolling at Lincoln Square in Chicago...we entered a book store..." Raven's Woods" (it wasn't planned)...All kinds of books...awesome variety of books...It wasn't a big shop but the concentration of books per square metre of the floor area was the highest I had seen so far...

The owner asked me, "Was I looking for something in particular?" I wasn't actually...so I asked him (thinking there was less than 1 percent chance that he would have it): "Do you have *Critique of Pure Reason* by Immanuel Kant?"

I was astonished when he said: "Yes I have it...you won't find it...I'll help you..."

Even he took about seven minutes to locate it...It was bound and in good condition...printed in 1955...it was the second edition...written in 1787 (first edition written in 1781)...price $8.00...I tell you...It was a steal.

Dave Astor in reply to: SKSagar

SK, what a great story about how you got the Kant book! I love bookstores — like the Chicago one you mention — with eclectic collections, knowledgeable owners, and "ambience."

SKSagar in reply to: Dave Astor

Now...I have just finished my fifth reading of the paragraph in question...Believe it or not, I am beginning to understand it... and "Bion." It's nothing short of "spellbinding."

This is what I think, is the gist of what it says:

"Before we make our judgment of whether something is 'good' or 'evil,' we must ask ourselves under which law we are making the judgment...whether it is the law of freedom (of will)...or it is a physical law (law of physics). The former from considerations of 'pure practical reason' corresponds to the conduct of beings in the 'world of intelligence,' and the latter from considerations of pure reason (theoretical) would mean that any act performed is in the world of sense, i.e., just a phenomena, i.e., merely the motion of atoms and molecules in accordance with the laws of causation."

Perhaps if I read it five more times with ever increasing levels of concentration...I may understand it still better and realize that every word of it is in place.

And to think...there are about 1,800 such paragraphs in the book, and if you want me to support your statement that "Some difficult books are not as difficult as we think" in respect of Kant's book...I am afraid...life is too short.

Will come back...after reading five more time

Dave Astor in reply to: SKSagar:

SK, it's very impressive for you to have read the paragraph five times in order to finally start to understand it! I might have given up after two or three tries.:-)

After reading your "translation," and rereading the Kant paragraph, I think your take on it makes sense, which leads one to wonder why Kant couldn't have written that paragraph (and book) in a less complex way...

I also enjoyed your humor at the end of the comment!

Conversation ends.

Later in life I intend reading this book. Whether I'll be able to do so or not remains to be seen...but there is no doubt this is a priceless possession...(next only to Erwin Schroedinger's *What is Life?*).

I happened to read the bibliographical note and learned that it took him nine years to complete the book though he declared that the actual time he took in writing the manuscript was between four and five months. Is that possible? I think not...I think it is utterly, utterly impossible to do so. Unless..."There was someone in his head...and it wasn't him."

And as he took nine years...it is possible..."That someone" left him intermittently, and kept coming back.

6-015:

And then it happened...An NDE ensued last night...the second NDE of the calendar year...sometimes I feel it is risky to sleep at night.

Lightning struck...for a second or two...it was like an electric shock of some severe magnitude...Perhaps it was a dream... but the degree of consciousness wasn't corresponding to just a dream...motion of atoms and molecules in time and space...inside the body was felt in a distinctive manner...enough to ask the question...did I wake you up? Did you hear a sound? "No" was the answer..."Now you did."

What caused it...was it a false creation, proceeding from the heat oppressed brain (Macbeth of Shakespeare)? Was I thinking about a certain correlation...a certain entanglement...Was it the same "something or someone" who visited Kant...intermittently...and enabled him to write that awesome book? Was it the same "sos" that visited President Kennedy and supplied him with a certain strength of character...a certain restraint...and a certain stunning presence of mind that enabled him to reduce that index of probability of a nuclear holocaust (the Cuban missile phenomena)? Was it the same "sos" that visited Malala, that fourteen-year-old Pakistani girl, and supplied her with such courage and wisdom and such intensity when she spoke those words...words that may change — for the better — the mindset of her countrymen and make it more favorably inclined toward establishing gender equality... and so on...innumerable examples can be given.

And then…enlightening struck…Perhaps "it" came back for a short while…and supplied me with all the inputs I needed to compose my talk for the third day of the seminar.

**

Chapter Six C

DAY THREE

CONVERGENCE TO "SOMETHING"

As usual, on all days, recorded classical music was in the air, in the auditorium for the first fifteen to twenty minutes while the delegates were arriving and settling down in their seats. Even when the speakers began their talk, the music went on albeit at a low volume. On this the third morning of the seminar, it was Grieg's Piano Concerto in A sharp minor that was going on when the MOC came on the stage to begin the day's proceedings by inviting the next speaker. She seemed to act strange, taking her time to start talking; a certain portion of the music was of such profound beauty, it was beyond her capacity to ignore it. It looked odd, but the crowd understood, and even applauded the silence, when that portion was over and the volume was finally reduced. It was a unique atmosphere, and the stage was set for another interesting day at the seminar.

The MOC requested "yours truly," hereinafter called "SK," to come forward with his presentation to throw some light on what these "six words" were all about.

Amid all this some famous quotations of scientists and philosophers, including some statements/deductions by SK (6-016) were being continuously displayed on the screen. These were also shown in the seminar catalogs made available to all the delegates — and the six words could be decoded from them.

SK (or perhaps someone who played his part) then slowly got up from where he was sitting, a picture of nervousness — by no means used to be in the midst of such an august gathering — even tripped and tilted slightly, but composed himself somewhat and reached the center of attention. Still nervous — he looked at Stephen Jay Gould's revised version in the dais, and remembered the maxim (from the chapter "The Streak of Streaks" of SJG's *The Richness of Life*) "When you're hot, you're hot; and when you're not, you're not." He must not get rattled, his hands should not start shaking, he must get the touch, build confidence, find the rhythm...swish, swish, swish...and all nervousness faded. After paying due respects to the concerned many, and after highlighting his own standing as a common man (with a certain configuration and quality of the brain cells supplied to him by the god who plays dice) in the midst of geniuses, but a concerned man nevertheless, began his presentation thus:

"Before I let you on to the 'six words' (which I presume most of you would have guessed — by decoding the quotations — already) let me go back to John Wheeler's story of his experience of a certain twenty-questions game played by him. This also I presume is known to many of you. In the words of Wheeler himself:

"We had been playing the familiar game of twenty questions. Then my turn came, fourth to be sent from the room, so that Lother Nordheim's other fifteen after-dinner guests could consult in secret and agree on a difficult word. I was locked out unbelievably long. On finally being readmitted, I found a smile on everyone's face, sign of a joke or a plot. I nevertheless started my attempt to find the word. "Is it animal?" "No." "Is it mineral?" "Yes." "Is it Green?" "No." Is it white?" "Yes." These answers came quickly. Then the questions began to take longer in the answering. It was strange. All I wanted from my friends was a simple yes or no, no or yes, before responding. Finally, I felt I was getting hot on the trail that the word might be "cloud." I knew I was allowed only one chance at the final word. I ventured it: "Is it cloud?" "Yes," came the reply, and everyone burst out laughing. They explained to me there had been no word in the room. They had agreed not to agree on a word. Each one questioned could answer as he pleased — with the one requirement that he should have a word in mind compatible with his own response and all that had gone before. Otherwise, if I challenged, he lost. The surprise version of the game of twenty questions was therefore as difficult for my colleagues as it was for me."

Coming back to the six words, I ask the question:

"What can be the six words, which find compatibility with all that has been said before in this seminar? — compatibility in the sense of either "in agreement," or at least "in support" of the conclusions drawn such as:

By Einstein in his talk on "cosmic religious feeling"

By Schrödinger in his talk on "oneness of mind" and the famous equation of the Upanishads. The "Flickering light thought experiment," the "indestructibility of mind and its peculiar time table that it is always 'now,'" also of Schrödinger.

The "universe as a self-excited circuit," the "participating observer" of John Wheeler.

The consciousness philosophy of Wigner that "consciousness of the mind can be the hidden variable that explains some of the features of quantum mechanics."

The "transactional interpretation of quantum mechanics" of John Cramer, the "quantum entanglement phenomena," and conformity with Einstein's special relativity *vis-à-vis* the "length contraction" aspect.

By the philosophy of Descartes and the statement: "I think therefore I am."

By Kant in his reasoning on "the unity of consciousness."

By de Reincourt in his talk on "squaring the circle — beyond the mind."

It must be understood that the "six words" do not deny the existence of a creator. On the subject of the existence of a creator, it is imperative that a scientist's role must be understood properly and in this respect the scientist, the philosopher, and the theologian must have a common ground. The contention of some theologians that the question of the existence of a creator is by "definition" outside science's domain is self-contradictory in the sense that the word "definition," which "defines something" or says "something is definitely true" is all in the center stage of a scientist's world. In essence, the theologian, who is made up of the same quantum entities as a scientist, has himself tried to make a scientific statement as it is based on certain observations — the observations being that "scientists have so far neither been able to prove the existence of God nor disprove God's existence." But this universe of ours, though nearly fourteen billion years old, is considered still in its infancy and there is ample time to search for the ultimate truth, and regardless of how long it takes, whenever these things are understood properly, they will have to be considered in the domain of science. If not, then we must answer the question "Whose domain is it?"

Besides, a majority of scientists do not question the existence of God, it's just that they cannot decide in what form He exists. My own view is that the expectation of a classical-level God made up of atoms and molecules and having mass leads us to a state of mind where we have no choice but to ask the question, "Are we in good hands?" Can such a classical-level God be considered omnipotent or omnipresent? How can this be possible? He would have to occupy a certain limited region of time and space, however large it may be, and be subject to interactions and forces caused by the world outside this region, on which He can have no control whatsoever unless He breaks the laws of physics created by Him. If He sits in judgment in a limited region of space, in say, the Andromeda Galaxy and keeps a comprehensive record of the deeds of all beings on the planet Earth or any other planet in any other part of the universe and then rewards and punishes accordingly, he will have to violate Einstein's special theory of relativity and Newton's laws of motion (at the classical level) and many other laws of physics all created by him, in a continuous manner. The trouble is, we don't see this happening. To think that such a classical-level god will come to our rescue and prevent the civilization from self-destruction, will be wishful thinking in the extreme. On the other hand, the expectation of a quantum-level God—who is not required to follow the laws of classical physics—and who is conscious and omnipresent in the form of a cosmic consciousness, having created the laws of physics, including the laws of probability and most importantly the "anthropic principle" is easier to believe. But, in that case, Roger Penrose's statement given below must be brought to mind.

"Probabilities do not arise at the minute quantum level of particles, atoms or molecules—these evolve deterministically—but seemingly via some mysterious larger scale action connected with the emergence of a classical world that we can consciously perceive." (*The Emperor's New Mind*)

The conclusion is inescapable: There is no other way to prevent such a collapse of civilization from self-destruction except to create such probabilities at the classical level so as to enable the quantum-level God to come to our rescue. And the only way these classical-level probabilities can be created is a scientific approach to the understanding of reality and to base our religious beliefs on science and science alone, which calls for a comprehensive convergence of science and religion, not to mention a comprehensive convergence of all the religions into one religion.

In the eyes of the quantum-level God as an observer, he sees the scientist and the religious leader as made up of the same quantum entities, atoms and molecules, who could easily have changed places, had their interactions been different. In principle he sees no conflict at all between the scientist and the religious leader. Their roles in understanding reality are complimentary. The former tries to answer the question—"How we are here?"—by experimental observations and applying the rules of science (physics and mathematics), and the latter tries to answer the question—"Why we are here?"—by intuitive thinking, by taking into consideration the updated status of science, making sure that all the scientific truths acquired and established until that time are not rejected, while at the same time firmly believing in the existence of God until such time as (and even after that) science is capable of supplying answers to the ultimate questions about why things exist and what is their purpose.

Under the circumstances, it will be the wisest thing to order our lives on the assumption that God exists at the quantum level, but spread over the entire universe, and creates the laws of science, in a way that He cannot violate these laws himself, and these laws include the law of probability, which too He cannot violate, for he too can just fix probabilities and cannot predict the outcome of an event with certainty, and yet he does all this with a sleight of His hand in a way that leads toward a consciousness that can understand His existence as well as the existence of the universe, and all this He manages to do (the sleight-of-hand part) by creating a certain principle that provides certain mathematics constants to all the laws of science, resulting in full obedience to all the laws, including the law of probability, and it is this principle, which we call the anthropic principle.

In short it will do no harm to anyone's religion, to believe firmly that God exists and He is within us and outside us, and that God is a scientist who created certain principles and laws that enabled the existence of life that understands the universe, and believing this is our religion and everybody's religion. There are no Hindus, no Muslims, no Christians. There are no Indians, no Pakistanis, no Chinese, no Americans. And there are no Democrats, no Republicans.

**

From the audience: What then are these six words?

In a low voice SK uttered the words, which were also displayed on the screen for just one second:

**

(6-017)

And then he spoke the ten words from Erwin Schrödinger's *What is Life?* and the same were also displayed on the screen:

**

And finally the equation from the Upanishads was also displayed on the screen:

**

THE EXPLANATION

The omnipresent "mind" is at the center stage of the philosophy of the "six words."

What is implied by the word "philosophy," and what do we understand about "philosophy," "Religion," and "Science?"

Guess I'll use a little algebra for this purpose:

If x corresponds to the ultimate truth that is to be arrived at (may not be possible before a few more billions of years), and y corresponds to the portion of truth that is fully understood and established by science, then $z = x - y$ obviously corresponds to that portion of truth, which is either unknown or partly known and not yet established by science. The entire philosophy of the world — whether written in books/websites, or discussed in seminars/get togethers — is dealing with z, and that portion of it, which is related to such topics as "mind," "consciousness," "soul," "spirit," and "God," etc., called "religion." As and when a portion of z becomes an established truth by science (by experiment/observation), it gets added to y. Until such time, it remains in the domain of philosophy/religion. In my opinion the basic fundamental criteria to be adopted in arriving at one's religion is that it should be fully compatible with y and should not be required to reject an already established truth.

Let us now discuss the z category subject of the "omnipresent mind" in relation to the already established truths y pertaining to the following:

"Time or shall we say pseudotime," "Space or shall we say pseudospace" "Quantum entanglement," "Bell's Inequality," "Einstein's special relativity," "The collapse of the wave function," "The participating observer," "The probability wave," etc.:

Two z category assumptions are contemplated here:

The first assumption is that this "mind" of ours has the capacity to "effectively stand outside of time" and study this phenomena called "time."

The second assumption is that this "mind" of ours has the capacity to "effectively stand outside of space" and study the phenomena called "space."

Consider that I—call me "A2"—while giving this talk right now at this very moment decide to look outside the window toward the sky and start gazing at a galaxy. My friends tell me that it is such and such and it is seven billion light-years away. In that galaxy there is a star like our sun—I am looking at it too. Within the planetary system of that star, there is a planet like our Earth—I am looking at it too. On that planet there is a man called "A1" and he too is looking outside his window. In principle no one can say that I am not looking at him now. I see him now, but according to him, the universe is only seven billion years old, whereas according to me it is fourteen billion years old (assumption), but we are in the same universe. Astonishing, is it not? But, of course "A1" knows the truth; he is a cosmologist. In principle, it is also possible he is now thinking about me, as I am now starting to think about the man called "A3" on a planet orbiting a star in a galaxy, which is fourteen billion light-years away, and it is he who is looking at me, and his universe is twenty-eight billion light-years old. I cannot see him as he is in my future. If I am really able to look that far out from my window I should then be looking at the Big Bang as it unfolds, same as A1 if he were looking as far out from his window as he is from me. And then there is A4 whose universe is fifty-six billion years old, A5 whose universe is 112 billion years old, A6 (224 billion yrs), A7 (448 billion years) and so on and so forth. So, it seems different people living in different galaxies scattered far apart from each other think the age of the universe as different from each other. We are just a select few who think the universe to be fourteen billion years old. Are these guys called A5, A6, A7, more knowledgeable than us? They should be, but I doubt it very much, particularly on matters of cosmology. Consider for example A20, we A2s over here think A20's universe to be 3670.016 trillion years old, but does A20 think like that? At that point of time after the Big Bang, the galaxies are so far apart, they must be completely out of sight of each other even with the most powerful telescopes available to them. When A20 looks out of his window, he cannot see any other galaxy at all; all he can see are stars from his own galaxy, and so he will think that's all there is to that for his universe. It looks as if we A2s know much more about the universe than the highly experienced A20s, and it's possible that A1s are the most knowledgeable of all.

Let's now talk about the "mind." Now this "mind" is standing effectively outside of time and contemplating all this. The mind has no problem whatsoever in arranging this connectivity. All these guys A1, A2, A3, A4, A5, up to at least A10, are now looking outside their windows and thinking about each other. They have studied quantum physics and cosmology and understand what's happening. How can all this be possible unless it is the same mind everywhere? Perhaps A20's knowledge may not be up to the required level, but the mind inside his body is well aware of all this for the simple reason that it is the same mind. And so this good old mind of ours is not only omnipresent in space but also in time.

And, in principle, it explains quantum entanglement. The mere fact that there is such a thing as quantum entanglement implies instant connectivity, and this connectivity is impossible to imagine except by consideration of an omnipresent mind.

What exactly happens in entanglement in relation to this omnipresent mind?

When two quantum particles — once in contact and now far apart — entangle once again, could it be that it's the mind that does the traveling to and fro? It does it instantaneously and facilitates the connectivity, perhaps — as in John Cramer's transactional interpretation based on the Wheeler-Feynman emitter-absorber theory, it takes the quantum particle with it in the form of a retarded wave that goes into the future, and after the said quantum particle's shaking hands with its old friend and completing the transaction, returns back into the past as an advanced wave, the former journey at speed of light and the return journey at negative speed of light, thus taking no time at all.

Now, let us bring in the intelligent as well as the participating observer. He has two categories of intelligences built in inside of him: 1) the permanent and the all intelligent mind. Let's call it "M," and 2) the temporary and the acquired intelligence — through the interactions of the world — *vide* consciousness. Let's call it "C."

When this mind "M" is engulfed in "matter" of certain proportions through the medium of a certain biochemistry, the resulting consciousness "C" — with the help of interactions of the world that impart information and intelligence (the temporary type) to it — can as per Wigner's hypothesis "collapse the wave functions" whenever it makes an observation, and it is this consciousness that enables the observation that sees a reduction in the length of an object when the observer is traveling at close to the speed of light. But, for the mind itself, this requirement that it should be engulfed in a consciousness in order to collapse the wave functions is not a necessity. It obeys Einstein's special relativity, and travels at the speed of light and nothing more than that, and measures the distance between objects — however far apart they happen to be — as nil, which explains its omnipresence.

The trick that the God who plays dice has played on us with the sleight of His hand is understood. He allows the mind to travel at the speed of light but not the conscious container of the mind, and further He allows the mind that travels at the speed of light the power to collapse the wave functions and to measure the distance between objects as nil. But, the same powers are not provided to the information that these conscious containers may have sent, even though the said information could travel at the speed of light.

Consider now the second assumption that the mind can effectively stand outside of "space" and study these phenomena called "space":

If the mind does so, it will be able to see what exactly the shape of the universe is. Only someone who can look from outside the universe at "cosmic time—which means the current age of the universe at all locations—about 13.7 billion years or so—can understand what is the shape. To me it appears that the universe is perfectly homogeneous and I am at the center. To someone who is in the Andromeda Galaxy, it will appear the same way that he or she is at its center. But the trouble is none of us can look at it, when it is "cosmic time" at all locations, for we can only look backward in time. When I look at the evening star "Sirius," which is about eight light-years away, I see it as it was eight years before. When I look at another galaxy, I am seeing it when it was several billions of years younger than me—assuming I am 13.7 billion years old—where is that galaxy at precisely the present moment is impossible to say. In general, there are three kinds of motion—after the initial inflation—that determine the shape: dilation, shear, and rotation. There can be millions of permutations and combinations with regard to the proportions shared by each category of motion. Finally, what is adopted—or what was in the mind of—by the Super Intellect—(will talk about it later)—at the time of programming the new universe—might have been a hugely complex combination of these, which gives the best results. What can be the parameter or yardstick selected when arriving at the best result? Perhaps it might be to create the longest time before the galaxies collide with each other.

But the mind sees it all not just at the "cosmic time" pertaining to the "now" of ours, but also at the "cosmic time" pertaining to the "now" of "A1," not forgetting "A3," "A4," etc. It is there at the "cosmic time" pertaining to the "now" of someone who was there at Planck time after time zero of our universe, and it is there at the "cosmic time" pertaining to the "now" of someone who will be there at Planck time before time maximum of this universe, which will coincide with Planck time after time zero of the next universe.

And invariably the "mind" finds a biochemistry...interacts with it...and attains "consciousness"...And these two guys...mind and consciousness are forever playing tricks with the universe.

To explain what I mean...Let's go back to Wheeler's conversation with Wigner and Wheeler's statement: "One has to recognize right from the start that in all the theory of black hole physics that we do today...We feed in the assumption about the direction of time—we do not deduce it. We put it in as an initial condition. We could perfectly well talk of holes that work the other way round, just as we could perfectly well talk of heat going from a cold teacup to a hot teacup. But it would violate all our understanding that it is initial conditions that count and not final conditions.``

I ask the question, "Who is *we* here? "Who" was it—in the literal sense—that was responsible for setting these initial conditions? Was it not "consciousness" that played tricks?

Wigner's Next Question: ````In thermodynamics we say, when deriving the law of entropy that the initial conditions are irregular except for those that are macroscopically constrained. Otherwise initial conditions are fully irregular. This is the basis of the increase of entropy. What is a similar postulate for the black hole?"

Let's discuss the second law of thermodynamics:

A glass on the edge of the table falls down and is broken into pieces...we say that "disorder" (I use Boltzman's term and not "entropy.") has increased toward increase in time...time cannot be reversed back in the sense that the pieces of glass on their own cannot reassemble and climb back up on the table.

The glass was in a state of "order" while it was on the table and then it fell down and the "disorder" started increasing as per the second law.

Now, consider the earlier state of the glass...while it was being manufactured in an industry...obviously it was less orderly and the "disorder" kept changing/reducing until it was completely ready and placed on the table...

Now you might say the system itself was changed.

My second question: How is it that human beings — who are themselves part of nature, being made up of trillions of constituent parts such as atoms and molecules — can alter the state of the system from "closed" to "open" or "open to "close" as they like and thus alter entropy? And that they do it without changing the direction of time.

In short is it not *consciousness* that is playing some tricks?

Before my next question...Let's discuss the direction of time. Now the direction of time cannot be changed...at least for the duration of this Universe...

This was after all the initial condition established by "The Consciousness" (rather the Super Consciousness that programmed and then switched on the Big Bang).

Even in the case of a Big Crunch, the direction of time will not change...But first...Can there be a Big Crunch?

Let us look into the deep future...trillions of years deep. Current levels of knowledge say that expansion will continue forever that the dark energy will keep driving the universe toward perpetual nothingness.

But there is one scenario we need to consider. That after a great spell of time the galaxies, while moving away from each other, will get so far apart that they will have no influence on each other. Each will be like a universe in itself and matter density will be enough for gravity to overpower entropy...and the galaxies will start collapsing on themselves toward their black holes at the centers...

Slow at first and gaining speed in course of time...It could still take billions of years. That could be the Big Crunch...But the direction of time will still be positive toward the future...The glass that falls from the table will still be moving toward disorder... it will still not be able to reassemble on its own and climb back on the table. It's just that the orbit of the solar system around the center of the galaxy will start reducing...

Only near the end will the situation become critical and dangerous. But, by then, the degrees of consciousnesses of the beings of the stars and the intelligence levels would have increased a thousand fold (or maybe a million fold…what do I know?), enough to come together and form a Super Consciousness and then in time feed in the constants and press the appropriate buttons and at least achieve a somewhat controlled explosion of the black hole.

In fact the final question asked by Wigner to Wheeler near the end of the dialogue was this:

WIGNER: I'll ask one more question, then I'll shut up — Is it possible to imagine the explosion of a black hole? I often think that the Big Bang was an explosion of a black hole. Is that nonsense?

WHEELER: Zeldovich and Novikov proposed it some years ago — the idea that there are in addition to black holes, white holes. And that they should vomit forth matter the way a black hole sucks in matter…

Of course Wheeler's reply…in continuation…was not encouraging toward that conclusion (that a black hole explosion could cause the Big Bang)…but than a Super Consciousness a thousand times more intelligent will know what to do.

And so I come to my next question, which is, in fact, similar to the first two: Is it not *consciousness* that does the trick…and keeps doing all the time?

The hoarseness in SK's voice was substantially evident. MOC announced a short tea break. When the audience assembled once again, the MOC asked the delegates if they had any questions or views on what was spoken until now. Several hands were raised and a hugely interesting dialogue ensued. (6 -018)

C: I must caution you SK that just because the universe has many symmetries does not give rise to the notion that the universe is in fact a simulation.

DP: Very good C…

SK: But the Universe has many constants—extremely fine tuned—embedded into the laws the universe is following. That caused theoretical physicists to arrive and then understand the universe...which gives rise to the notion that the universe is in fact a simulation.

CFT: The idea that the universe is a "simulation" is not even scientifically coherent. Let us say that fine-tuning of natural laws is proof that a universe, which has those laws, is a simulation.

The universe in which our Universe is being simulated must be at least as "mathematically powerful" as the universe we reside in, because otherwise a simulation of our Universe would not be possible. There are vastly more conceivable worlds, which are less "mathematically powerful" than our world; therefore any world, which can simulate our world, must also be fine-tuned.

If the world that simulates our world is fine-tuned, then that is proof that it is *also* a simulation. Therefore, there is a third world, which is simulating that one. By the above logic that world must be simulated, and so on.

Therefore, by your logic SK, no universe is nonsimulated, so there is no meaningful distinction between a simulated and a nonsimulated universe.

In fact, a sufficiently powerful simulation is mathematically identical to the "real" thing, so the idea of a simulated, as opposed to nonsimulated, universe is entirely incoherent.

DOA: It's not that simple...CFT...If it were that simple, it would be easy to rule out the Cantor set. Whether something is coherent does not (NOT) depend on whether it is "scientifically" coherent. To the contrary: it is the job of science and expertise and education to make sure that at the very least "coherence" is not a property that needs to be judged by experts.

And given the Cantor set that's certainly not easy. The difference between pathological and non-pathological arguments should not itself be subject to debate.

CFT: We can rule out the Cantor set...*as an object in a physically realizable universe.* But that's beside the point. My argument was using *reductio ad absurdum* on the *specific* argument used by SK that fine-tuning of laws somehow represented evidence of a *physical* simulation of our Universe in another universe. There's a whole host of other, thornier philosophical arguments to be had about simulations in general.

I don't think the idea of our universe being a simulation is coherent, but I should have phrased it better to avoid making it seem as if that was the intention of my argument, when the argument was merely that saying that fine-tuned laws was evidence that the universe was physically simulated in another universe was incoherent.

SK: Here we are talking "philosophy"...Neither can I say with a 100-percent degree of certainty that the universe is a simulation...nor can anyone say with a 100-percent degree of certainty that it is not a simulation...

This universe, though 13.7 billion years old, is considered still in its infancy...The degrees of consciousnesses and the levels of intelligence are forever increasing and there will be ample time to know the truth regarding "simulation."

For the time being let's consider the evidence available and then judge for ourselves what is the probability that the universe could in fact have been "simulated"...

And when we do that...When we study the evidences available...When we look at these constants...When we see how extremely fine-tuned they are in such an extremely narrow range...And there are so many of them, perhaps more than twenty...And each one of them has to be that finely tuned for the desired end results...We will realize that the probability of the universe not a "simulated" one is just about as low — as someone said — as that of an aircraft getting assembled by a tornado striking a junkyard.

CFT: That's not a good (6-019) argument SK...All teleological arguments about the nature of the universe are no good (6-019). The argument from fine-tuning, which you are using makes the assumptions that:

A.) Natural Constants are 100 percent arbitrary (i.e., they could just as well have been any value, with intrinsic changes to our physics)

B.) Life of *any kind* must be like ours and can only exist in "possible" worlds very close to ours.

C.) There is only one universe and it is this exact one we are in.

First, you blow C out of the water by making this an argument about simulation. If there's another universe where we're being simulated, maybe there are an infinite number of universes with slightly different laws. You have no reason to pick "simulation" over "unguided anthropic selection." It's a matter of choice. What we know is perfectly consistent with a multiverse generally inhospitable to life.

With B, there are surely an infinite variety of information processing species possible in all of these infinite "possible" worlds, according to their physics. Basing your argument on the necessity of carbon-based Earth life "isn't good logic"(6-20).

Regarding A, we have no reason to assume the natural constants you think are "fine-tuned" are arbitrary. For all we know they can be established mathematically from fundamental principles. Choosing to believe that they are arbitrary is, again, a choice, and a totally unscientific one.

SK: Maybe...The required quantum of simulation is not that high...It is not that each and every activity is a consequence of that simulation...Perhaps it is somewhat of a notional simulation. Only a computer program was prepared and the mathematical constants provided as inputs by the designers...But the designers had no control over "randomness," "probability," and the "laws of causation."

The designers could not say when and where life would emerge and then understand the universe...or when a civilization from planet x would conquer a civilization from planet y... or if Obama would win a certain election. One just presses a button and then sees what happens...

This also explains all the imperfections in our world. It's like when we press a button to cause a massive explosion in a crowded city. We are sure the damage will be done...but we cannot say which buildings will be destroyed or precisely how many people will be killed.

CFT: The "imperfections" of the world are a well-understood result of chaotic dependence on initial conditions and thermodynamic entropy. They don't need a metaphysical explanation beyond that.

And as for your point of view that "It is not that each and every activity is a consequence of that simulation...Only a computer program was prepared and the mathematical constants provided as inputs by the designers..." the problem is that you need some computational substrate for this simulation to exist in, so in fact every activity *has* to be a consequence of that simulation. What I mean is you can't just say (to be flip) "Let there be light" and have a simulation of light appear; you have to have a system that can simulate it. That's what a simulation *is*. The point I'm making is that *you* seem to be implying that the simulation is a mechanism being controlled by designers. My argument is that the computational substrate could as easily be an abstract simulation in the realm of Platonic forms. All simulations of our universe would be equivalent mathematically, so there's no need to assume more than necessary, like that there are intelligent designers behind its creation.

SK: CFT999...(I like the name) With regard to your argument... "That the mere fact that there are those fine-tuned constants embedded in nature is no evidence that the Universe was 'physically' simulated by someone in another universe"...This indeed is a valid point...and here again we are talking philosophy...rather deep "philosophy." Neither can I say with a 100-percent degree of certainty that the "simulation" was actually a "physical simulation" carried out by the Superconsciousness of the pre–Big Bang universe, nor can anyone say with a 100-percent degree of certainty that the "simulation" happened on its own...i.e., by "magic" or something.

Both alternatives require "carrying a quantum of metaphysical baggage." Here, I think it can be said that in the former case, i.e., "Super Consciousness doing the trick," it's a finite quantum of metaphysical baggage that needs to be carried, whereas in the latter case, i.e., "magic doing the trick," it looks like this quantum is enormously high, perhaps infinite.

CFT: It is rather deep philosophy. I tend to side with Plato on the subject—specifically that the universe we live in is a partial representation of a perfect mathematics reality. Some might look at this and call it "simulation," and that phrase has been used in Plato's presence an awful lot.

However, my major point is that there's a huge difference between the Platonic simulation (matter simulating form) and computational simulation (form simulating matter), which is what most people are talking about when they argue that the world is simulated.

Both have some concept they are modeling and some substrate they are using to model it with. Plato's forms are pure mathematics ideas, and the substrate is matter—essentially, matter acts as the computer and the forms the software. A computational simulation (like the ancestor simulations some trans-humanists talk about) has a preexisting idea of reality as its model and some kind of computer as its substrate.

The huge difference here is that in the Platonic simulation, our physical reality is a byproduct of the simulation of forms. In the computational simulation, physical reality is the intended end product, which brings into question where the physical reality being modeled comes from and leads to infinite regress.

DOA to CFT: The argument in favor of our universe being a simulation is a different one (6 -020a):

It's core goes like this, and it's perfectly plausible — as far as it goes:

"A long-proposed thought experiment, put forward by both philosophers and popular culture, points out that any civilization of sufficient size and intelligence would eventually create a simulation universe if such a thing were possible. And since there would therefore be many more simulations (within simulations, within simulations) than real universes, it is therefore more likely than not that our world is artificial."

It's merely a challenge. Food for thought. But it's also good enough to debunk the entirety of New Atheism in one paragraph.

I am *so* sad that I didn't come up with that paragraph. But at least I'm not denying it.

If you just take this idea and ask yourself: What would happen to folks living 5,000 years ago when they came across something like it? What would they do? Prove that the continuum hypothesis is independent from the remaining axioms of Zermelo Fraenkel set theory? Or come up with a religious myth of the creation of the world by a superior intelligence?

And most importantly, which one of the two has better chances to survive for the number of millenia it takes so that somebody can tackle the foundations of mathematics — telling us that the answer is approximately 42.

CFT: Uh huh. You're talking about ancestor simulations; but, SK, I was responding to using a fine-tuning argument, so that's what I focused on.

The problem with this idea is that we have very good reason to think that it is *not* possible to create such a simulation, except on a limited scale. That is to say that fundamental limitations (in our case the Bekenstein bound, which limits the amount of information that can exist in a volume of space, but also thermodynamic considerations, not to mention the economic impracticality of such an endeavor) would force our simulation to be limited compared to our universe—or, in other words, a simulation of the universe utilizing all the space and energy of that universe would *be* that universe and not a simulation.

Therefore, because all simulations are small, and nested ones even smaller, the original universe would be staggeringly large and contain the vast majority of people. This, I think, totally invalidates that argument.

DOA: So, you rely on the strong Leibniz principle and a rough estimate on sizes of abstractly possible universes in order to settle a metaphysical dispute that is at least as undecidable with finitist means as the fine structure of the Cantor set?

Some people are really brave around here. And as for your remark on SK's argument regarding "all teleological arguments about the nature of the universe are stupid," that's a pretty strong claim given that the best theories about the nature of the universe are so far unable to reconcile the very small and the very big. Under those circumstances, who is to rule out the possibility that the only way to "round up" our understanding may even *require* teleology? Or something similar, like entelechy?

Because you don't like theology and metaphysics? Is that the reason why? Then please explain to me the nonmetaphysical documentation of this:

"We have no reason to assume the natural constants you think are "fine-tuned" are arbitrary. For all we know they can be established mathematically from fundamental principles. Choosing to believe that they are arbitrary is, again, a choice, and a totally unscientific one."

First of all, "we" clearly *cannot* establish mathematically from fundamental principles the constants of the Standard Model. But even assuming that "we" could, how would that be nonmetaphysical and nontheological?

For all I can tell, you have *no* way of ruling out that the answer will prove equivalent to some variant of entelechy. In fact, the only alternative seems to admit that theory will arrive at the limit of its wits. Or has already done so.

CFT: "...how would that be non-metaphysical and non-theological?" You're using a straw man here. There's nothing inherently theological about mathematics. As for metaphysics, when you're talking about the mathematical underpinnings of nature you're talking about something *inherently* metaphysical. I'm not sure what your issue is here.

Teleological arguments are not good evidence of intelligent design. You can only ascribe them to intelligence when you've either:

A.) Given some other persuasive evidence that intelligence could exist (we have no evidence either way at this point)

B.) Proven that no other argument suffices to explain a phenomenon

In either case you need much more persuasive arguments to support the teleological one.

As for unguided teleology, like the strong anthropic principle, you still need evidence of some unguided mechanism, which can be selected for the desired outcome. Your argument may be true but it needs a lot of evidence.

So, simply, from the basis of logic, teleology should be avoided in science unless absolutely necessary. You're right that it may be true, but it's probably not.

As for establishing constants from first principles, we've done that already with some constants. Pi, for example, began as an observed phenomenon—every time we measured circles we got the same answer. But the value of pi arises essentially from the nature of Euclidian space—it couldn't be any other value. That's the kind of thing I suspect we'll find about the "fine-tuned" constants

DOA: It's very refreshing that you're not worried about metaphysics at all. Very rare too in these forums. And you clearly don't seem to be materialist either. We can quickly agree that indeed there is something inherently metaphysical about the role of mathematics in all sufficiently rich models of nature.

Whether that makes mathematics theological or not is another question. There are guesses and opinions though. For example, a famous Bourbakist once said that "We know that god must exist because mathematics is consistent, and we know that the devil must exist because we cannot prove it."

I didn't defend Intelligent Design (in the sense of the movement) and never will. Teleology and even Creationism is something else. What I claim is you can't prove them inconsistent—not with the means you need to assume when you believe in a unified theory of the universe, and not without those means. Hence never, no matter what.

Your optimism with respect to the mathematical meaning of the constants of the Standard Model is breathtaking, and on certain days, I would like to share that optimism. But then I watch the movie *Pi* and recall that numerology isn't science...

DOA TO SK: You're making very interesting points. But trying to answer too many things at once invariably leads to a lack of precision. And, ultimately, the assumption that there must be a way of making all of this stuff clear may not be warranted. I agree that we cannot rule out the idealist view that traces everything back to consciousness. But, historically, such attempts have always resulted in worldviews that lacked in resilience.

Scientism is merely the other extreme and doesn't work either. But the core problem is the lack of resilience. To avoid that problem, help only comes from pragmatism. And that means, among other things, accepting that we may not be able to figure it all out. Nor do we need to.

SK TO DOA: Regarding "help only comes from pragmatism. And that means, among other things, accepting that we may not be able to figure it all out," I already said that we may not be able to figure it all out...for a long, long time. But as the degrees of consciousnesses and levels of intelligence keep increasing with time, we will keep getting closer and closer to that—figuring it all out—stage.

Regarding "Nor do we need to," do we need to play football, watch movies, eat out, etcetera? We find it interesting and we do it for fun...

Besides philosophical discussions (not limited to the "simulation" issue) will forever remain a continuous process...forever moving toward a "world view" acceptable to people of all countries and all religions.

SK IN CONTINUATION: (In a lighter vein) Continuing the discussion on "Simulation issues"...In the broadest sense (or nonsense), there are only two players playing this game of "reality" and their names are: 1) "Matter and energy" hereinafter denoted by the symbol "MAE" and 2) "Mind and consciousness," call it "MAC"

Now there are two alternatives:

ALTERNATIVE 1:

MAE is the primary player—as most current day scientists would like to believe—and responsible for the occasional appearances of MAC at various locations and times, whenever—and wherever—it has "randomly" created a certain biochemistry for MAC to arrive.

In this alternative, there is no such thing as an anthropic principle and "randomness" is supreme and the kind of life, which is commensurate with the constants provided (also chosen randomly), materializes. This alternative does not explain how and from where does "intelligence" arrive on the scene. Perhaps it is an extremely rare commodity and also perhaps the quantum of intelligence is extremely limited, certainly not enough to prevent itself from self-destruction, and definitely not enough to avoid getting obliterated by a major asteroid hit.

Also, in this alternative, the universe in the deep future—with the kind help of dark energy (a major part of MAE)—will proceed toward nothingness...and that's it. No more universes—in that universe—no more MAE, let alone MAC.

Or if the galaxies in that universe at some time in the future start collapsing on themselves toward their black holes, perhaps new smaller universes will be created after the ensuing explosions. Randomness will still be supreme and randomly selected "funny" constants will determine what type of MACs will appear.

Lives will be rare in this alternative intelligent life; even rarer and whatever lives are there will be constantly killing and eating each other in an everlasting strife. If a choice is given to me I am not going to come back in this sort of universe. Now compare this with Alternative 2 where MAC is the primary player but has complete control over MAE.

Here in this alternative MAC on reaching a state of super consciousness, prepares a comprehensive program with beautifully designed mathematics constants and then switches on the computer at the appropriate time and ushers in the next Big Bang...and after trillions of years the next...and so on...cyclic phenomena.

An excellent universe with billions of galaxies and millions of stars in each galaxy, trillions of planets brimming with life and intelligence. Plenty of Einsteins and Newtons, plenty of Christs, and Buddhas too. Of course, randomness will be prominent in this alternative also, and so there will be some Hitlers too.

Here in this alternative I'll have no choice…I will be forever conscious (even if there are some time gaps between successive consciousness states for the simple reason that I will be unconscious of these "unconscious tenures," which will pass instantly for me)…

I can't help it…it is programmed that way. But it's nice… sometimes not so nice.

In this alternative you could call MAC as "mathematics" who gives orders and call MAE as physics who follows these orders. Indeed "mathematics" is the President and the company is called "The Universe."

And there are many such laws where mathematics is giving orders on how to proceed…and they were well crafted with nicely designed "constants" to create good-enough probabilities for lives and consciousnesses to evolve at millions of locations and theoretical physicists to arrive and then understand these laws: Laws such as the "uncertainty principle" where "mathematics" has played such a trick that it does not allow the quanta to get created out of nothing except for the shortest possible — and thus irrelevant — period of time. Laws such as "quantum entanglement" where "mathematics" has played such a trick that it permits entanglement but does not permit information to be sent faster than at speed of light.

And both these tricks have been played out by the mathematician by incorporating a certain randomness in the nature of reality. In the former case, the vacuum randomly fluctuates between being and nothingness and in the latter case the mathematician keeps shuffling the deck of nature in such a way that the randomness remains intact. I should go as far as to say that "randomeness" is the Vice President.

And then at the center of everything, there is this thing called "equivalence." The cleverest of all the laws, it goes without saying that this law is out and out mathematics and physics is just dancing to its tune. Who knows when — in the deep future — when dark energy drives the universe into perpetual nothingness…what will that energy get converted into…perhaps not into void…

In this alternative, good sense may prevail, science and religion may converge.. Conflicts of the world may be resolved, rise of the population may be reversed, and self-destruction avoided. And in this alternative, if an asteroid is on course to hit us, scientists may come up with a solution that saves us from a collision.

DOA: All right…I see…Pretty cool. I was just a little worried that maybe you weren't aware of the fact that speculative thought needs some constraints too.

Although I'm not entirely sure what exactly it would mean to "agree" with your point of view, I'm basically open to it. It might be the truth. But as I said, most of the time I prefer to act and think in contexts where it's just a little bit easier to tell whether or not I have a firm grip on the totality of propositions entailed by what I put forward (*aka* knowing the truth, according to Tarski). Which, together with the fact that it's an impossible feat, is the daily bread of the comedian and of the mathematician? But, of course, I would never go as far as claiming that this makes me a "mathematician."

It sure is good enough to link rational thought to "mimetics."

SK: Great…DOA. Yes, I am aware that speculative thought needs some constraints…and the biggest constraint is that the truths already established by science must not be violated…and I am trying my best to follow this principle. Your conversations with CFT was hugely interesting…it was..in fact…nothing short of spellbinding.

CFT indeed has some valid objections to "physical simulation by intelligent designers" such as 1) Some computational substrate is required for the "simulation" to exist in, and that this computational substrate could as easily be an abstract simulation in the realm of platonic forms, and 2) It is not possible to create a "simulation" except on a limited scale, a simulation of the universe utilizing all the space and energy of that universe would *be* that universe and not a simulation.

177

At the current level of knowledge it is difficult to give a satisfactory reply to counter these objections without recourse to some metaphysics or "teleology"...your views on "teleology/entelechy" too are very interesting and insightful.

But, then the current level of knowledge is indeed insignificant compared to what it would be a few thousand years from now. In some hundred thousand years we have become men from apes. It is reasonable to assume that after a similar time gap...or even less...we would become "Super minds" and then these problems of understanding "simulation" would not be insurmountable and what looks like teleology could then be nicely explained by science.

There can be several possible scenarios...of "physical simulations." A large-scale (full universe level) "simulation" carried out by the ultimate "superconsciousness" is a possibility that in no case can be ruled out. Erwin Schrödinger's (and Charles Sherrington's) concept of the "oneness of mind" together with a distinct possibility of the intelligent mind's omnipresence can lead toward the hypothesis of a Superconscious designer doing the "simulation." Perhaps it may not be "nonsense" to consider "intelligence" to be a compound...How can we—with our limited knowledge—deny this?

Perhaps the "omnipresent "mind" or "something" could also explain "quantum entanglement," not to mention "collapse of the wave function." Or, perhaps...I am out of my mind.

DOA: Regarding objections to physical simulations, let me admit that I don't believe at all in physical simulations of universes—and I don't think I ever claimed something else.

Also, I don't believe in extrapolating future knowledge or possible future knowledge. The use of the thought experiment elucidates aspects of our own evolution of understanding. I don't think it can help us find out something new about the world.

Sometimes it can be okay to violate "truths already established by science." But usually this requires the use of means that are extremely difficult to keep under control. To wit: the reason why some "truths already established by science" may turn out to be false is because those who established them were already in the same situation and didn't control enough of the consequences of their new approach, but nobody found out.

It's been like that for quite a while now. And to get beyond that dance (which is probably impossible) may require teleology or entelechy. But it's also possible that humans cannot include that mode of thought within a scientific worldview, which means that pragmatism may always remain one step ahead of theory? Why sacrifice the openness of the scientific process for the sake of the appearance of more comprehensive theory? The price is that the scientific worldview remains incomplete. But that's not a problem, as long as we don't expect from a scientific worldview what religion and philosophy promised but couldn't deliver.

As I said, we cannot rule out these scenarios (or some version of them). But that doesn't mean we can support them with evidence. It still means something that they can be made with some consistency: it means it's not surprising that people have come up with the notion and concept of a transcendent mind. There's nothing in these thoughts that violates science, or at least there is a way of rendering these thoughts that doesn't violate science.

SK: Perhaps intelligence "is" a compound...It's in the air...its omnipresent...Correlate this with the double slit experiment... the electron too is intelligent...perhaps. It is aware when it is under observation or not and does what it is expected to do.

Thank you DOA and CFT for your most valuable inputs

**

The interactions with CFT and DOA leads us to some conclusions:

We cannot say with any degree of certainty that our universe is a simulated one...nor can we say with any degree of certainty that it is not a simulated one, for there is no concrete evidence to support any of these two hypotheses.

I must follow a different approach to find an answer to the question "What is it that controls nature?"

I must place all the cards on the table…and discuss all the issues threadbare with an open mind…issues relating to the role of consciousness and that "something called "infinite mind"

The Role of Consciousness and the Infinite Mind

Some time back there was a news item that Stephen Hawking had launched the most powerful supercomputer in Europe. It's called "The COSMOS Supercomputer." It can open up new windows on the universe and unravel some of the deep secrets of the universe. It's obvious, of course that consciousness is at the back of this understanding that has led toward the development of the supercomputer…and will soon be leading toward an understanding of the secrets of the universe.

What role will the infinite mind play to ensure the continuity of life and consciousness in the universe? Let us place all our cards on the table, i.e., all the complications and the complexities that need to be addressed with each one of the various concepts such as the Big Bang, cosmological inflation, the second law of thermodynamics, the law of causation, the role of consciousness and the infinite mind, etc., proposed by the various scientists (or engineers…whatever) and let us discuss the issues threadbare with an open mind. (6-022)

Let's talk "inflation." The cosmological one: Was it the "vacuum energy" that caused a rapid increase in the size of a certain scale factor?

Now the scientist says, "The inflation of the scale factor meant that a small, smooth spatial region of the universe expanded exponentially to encompass a volume that would grow to become larger "today" than the size of the observable universe. In the process of expansion, the spatial geometry became flat." Reference *New Physics* edited by Gordon Fraser.

Consider the word "today" in the above statement. For us on the planet Earth "today" corresponds to about 13.7 billion years since the Big Bang and we say with our highly advanced —and at the same time highly limited — understanding of the subject... that sometime during the first second (between 10^{-36} secs and 10^{-32} secs to be precise), after time zero of the Big Bang, the universe inflated to a size (radius) of $10^{"z"}$ metres...where z is about twenty-eight.

Now imagine someone living on a certain planet orbiting a certain star in a certain galaxy, which is about 12.3 billion light-years away...consider that I am looking at him now...it's obvious that his universe is only about 1.4 billion years old. For that someone, if he or she is writing a book on cosmology "today" corresponds to only 1.4 billion years since the Big Bang...and in his description of "inflation" z will not be twenty-eight...rather it will be closer to twenty-seven.

In the same way for someone ahead of us in the deep future "z" will be greater than twenty-eight. What then was the true extent of the inflation with regard to the volume encompassed? Was Einstein's special relativity actually violated by the "entities" that traveled with the "inflation?"

Or, can we say that "special relativity — in respect of giving an upper limit to speed of light — " is not violated, in the same way as in quantum entanglement when we said that entanglement takes place instantaneously but no information can be sent faster than at the speed of light?

Or, is this value of z a measure of the consciousness of an observer as well as a measure of his coordinates in time and space at the time of his observation?

Or, is the concept of "inflation" itself misconceived? And, should we now look at some other explanation for the homogeneity of the universe?

Or, can we say that "There is indeed a physical law that 'prohibits' space expanding faster than the speed of light?"

Response from a delegate called LDM (6-021): "SK...You have raised many excellent questions. But I have a few points to make:

1) Most of the scientific community agrees that there was an early inflation of the universe, consistent with its enormous size

(observable part is about ninety-three billion light-years in diameter) versus its age (about 13.7 billion years).

2) The scientific community agrees a mass (or energy) cannot exceed the speed of light in a vacuum.

Based on the above, we have only two alternatives:

1) The laws of special relativity did not apply in the early universe, or

2) It was space that inflated faster than the speed of light.

No one really knows the answer. My judgment is that space expanded faster than the speed of light. This judgment is consistent with the accelerating universe we observe today."

SK: Thank you LDM…Let's talk about the Big Bang. For all practical purposes the Big Bang with all the activities associated with it—including inflation—is still in a state of superposition. Its wave function has not been collapsed as yet. Will it be collapsed any time in the future? I should think so…if it really was such a stupendous show—as it is made out to be by all and sundry—it could not remain a performance to empty stalls. The designers must have made sure that, in course of time, it will be witnessed by the consciousness of the observers of the future.

But it has not been observed so far.

So when you say that the size of the observable universe is about ninety-three billion light-years in diameter, what exactly is implied? Does it mean we can observe regions that are twenty billion light-years away, which means we can look back twenty billion years into the past…and en route somewhere we can pierce thru the Big Bang itself?

Surely the capacity of telescopes could not have been a constraint. We have observed supernovas that happened billions of years back in much detail—why not a few percentage points extra capacity and the Big Bang should have been observed clearly considering the enormity of its magnitude?

But the truth is it is not observed so far. What then can be the explanation? Perhaps:

a) The size of the observable universe is limited to the age of the universe — for the simple reason that when we observe something we are actually looking at its past — in terms of light-years, which works out as about 10^{28} meters radius.

b) Beyond this we cannot see no matter how great may be the capacity of our telescopes…as that would mean going back into time beyond "time zero" into a different universe altogether or maybe to a previous eon of our own universe…or maybe the main show of the Big Bang took place in a black hole from which light could not escape at all.

c) Or maybe the Big Bang was caused by some sort of a phase transition in the last stages of the previous eon of our Universe. This required a huge violation of the second law of thermodynamics in bringing about a comprehensive change in entropy from an infinitely high level to a near-zero level. Perhaps the Superconsciousness of the previous eon played a part in this transformation.

d) Perhaps the massless photons of the last stages of the previous eon were not getting bored at all…they were always up to something…getting together and forming a "Superconsciousness," designing the new universe with brand new and refined constants and then switching on the Big Bang.

LDM: "The Big Bang is the most accepted theory of the evolution of our Universe. It has some problems; however, it is not in a state of "superposition." It is widely accepted by the scientific community. And in Big Bang cosmology, the observable universe consists of the galaxies and other matter that can, in principle, be observed from Earth in the present day — because light (or other signals) from those objects has had time to reach the Earth since the beginning of the cosmological expansion."

SK: Appreciate your comment LDM...You are absolutely right... the Big Bang and inflation cosmology is very well understood and accepted by the scientific community...but it has not been physically observed — through a telescope that can look at a region 13.7 billion light-years away — nor adequately explained so far...As for inflation, it cannot be tested as the energies involved are too high and well beyond the experimental reach of our accelerators.

When I say the "Big Bang" has not been adequately explained...it is implied that an appropriate "physical cause" has not been determined (which is by and large acceptable to the scientific community) so far, of which the "effect" is precisely what happened at all the various stages of the Big Bang...not just the first second of it.

And that "physical cause," even if it happened just Plank time before the previous eon ended, must not violate any law of science...and it is further assumed that all the laws of science, which are applicable in the present universe, were also applicable in the previous eon.

Now what can be that possible "physical cause" that explains it? What can be that "hidden variable?"

Let's ask Eugene Wigner...then John Wheeler...then Rene Descartes...and then Erwin Schrödinger...and put all the cards on the table and see what consciousness and finally Superconsciousness can do to "switch off" one eon of the Universe and simultaneously "switch on" another:

Eugene Wigner's philosophy: It is the consciousness of the mind itself that acts as a hidden variable that explains the collapse of wave (phenomena).

In Wigner's own words:

"...the result of an observation modifies the wave function of a system. The modified wave function is, furthermore, in general, unpredictable before the impression gained at the interaction has entered our consciousness: it is the entering of an impression into our consciousness, which alters the wave function because it modifies our appraisal of the possibilities of different impressions, which we expect to receive in the future. It is at this point that the consciousness enters the theory unavoidably and unalterably."

John Wheeler goes one step further and feels the consciousnesses of the future can even collapse the wave functions of the past *vide* his concept of the "self-excited circuit," which in turn is deducted from his "delayed choice thought experiment."

Though I am an ardent fan of John Wheeler...I think this one is a bit too far-fetched...and difficult to comprehend...but this is a queer universe and anything is possible.

LDM: Again, you raise extremely insightful questions, and make excellent points. I'd like to add three points to your views.

First, there is no accepted theory as to what caused the Big Bang. I have put forth a hypothesis in my book, *Unraveling the Universe's Mysteries*. Second, we do not know what scientific laws applied at the energy levels associated with the Big Bang. As you pointed out, we are unable to duplicate those energy levels with today's science. Third, I discussed the problems with the Big Bang theory in my book and on my YouTube channel. The major issues are the initial inflation of the early universe and the almost complete absence of antimatter in the universe.

SK: Thank you LDM. Agree with the points made.

Regarding absence of antimatter...In my view this is a consequence of, one of the constants of nature provided at the time of programming by the Superconsciousness...that at the time of the Big Bang, one in about thirty million quarks would not find its antiquark and thus escape annihilation...which means all the matter available in the universe is made up of this surplus number of quarks...and it also ensures absence of antiquarks.

If this ratio had been higher—than one in about thirty million—matter density would have been higher and may not have permitted expansion of the universe.

Regarding inflation...this can be dispensed with...by assuming that the previous eon ended with a size that permits homogeneity...and by considering that the Big Bang was a phase transition that converted "something" into "temperature, radiation, energy, etc., whatever it was that came with the Big Bang.

All that remains to be answered is what was that "something?" What do you and I know? But that Superconsciousness knew...perhaps it was "itself." Now...I continue with the placing of the cards on the table on matters relating to consciousness and the mind:

Rene Descarte's views on the subject of mind and consciousness: "The mind is coupled to the body through the brain, which it uses (via the bodily senses) to acquire and store information about the world. It also uses the brain as a means to exercise its volitions, by interacting on the world. An important feature of this picture is that the mind is a thing, perhaps even more specifically, a substance. Not a physical substance, but a tenuous, elusive, ethereal sort of substance. It is neither perceptible by the senses nor extended in space; it is intelligent and purposive and its essential characteristic is thought or rather consciousness."

Erwin Schrödinger's views (including those of Charles Sherrington):

There is only one mind...But our consciousness is always in the singular...Now consider the billions of cells inside us... each one of which is a unit life in itself...and the awareness of each one of them is in the singular...But we the human beings are conscious of the sum total of those awarenesses...so there is a distinct possibility that there may "ultimately" be a cosmic consciousness, which is in the plural and contains the sum total of all the consciousnesses of the universe.

This leads toward consideration of the equation of the Upanishads: Atman = Brahman. Is it not mind boggling that this equation was conceived by the philosophers of the ancient past by intuition alone with no knowledge of quantum physics, Einstein's relativity, or even Newton's laws?

And we (some of us at least) of the twentieth and twenty-first centuries are in a position to talk about this equation, with the current level of our knowledge and place it "very much" in the domain of possibility. And surely the super minds of the future...say a hundred thousand years hence...will have enough knowledge at their disposal to validate (with some modification if necessary) the equation...even by experiment and observation.

Will come back to this equation...

But now, I must place the other category cards on the table: the many widely accepted scientific theories that predict the dark energy to lead the universe toward perpetual nothingness...no energy...no light...no life...no warmth...nothing but perpetual and irrevocable stagnation with massless photons and gravitons doing nothing at all—except getting bored.

And this goes on forever and ever...from age (say) 10^{18} years, to age (say) 10^{100+} years, since the Big Bang. Imagine...in such a scenario...that life and consciousness existing only in one unit of time out of nearly 10^{82} units of time. Does this make any sense at all?

Some scientists might ask, "Is making sense a requirement that the universe must consider?" And in such a scenario as described above...if "making sense" is not a requirement then life and consciousness was just an accident...There was never a designer...No anthropic principle...Just the second law of thermodynamics...Entropy will in the deep future be at its infinite maximum...there is no way...and no one...to change the direction...That's it.

Do we accept that? No we cannot...even to be aware of it, we need consciousness...and all that consciousness is concentrated in the tiny first region of time...up to the 10^{18} years stage...perhaps maximum up to the 10^{20} years stage.

And it is in the "here" and the "now"...while we are in this tiny region of time...while we are living in this Stelliferrous era... where beautiful stars are shining and beautiful life (with all that consciousness) is flourishing...while we with our consciousnesses are contemplating about the gloomy future (i.e., from 10^{20} to 10^{100+} years)...That the "infinite mind" must do something to avoid going into that region...but we need not panic...there is ample time...even in the Stelliferous era we are still in infancy.

What are the options available?...or...What are the possible scenarios? I should say that there are many...But they all need: The Equation: "Atman = Brahman"

"The personal self equals the omnipresent eternal self" (Schrödinger).

And that's a trump card I have placed on the table...And there is no teleology in this...no metaphysics...no law of established science is violated...some day science will deal with it... and explain it.

And they all need mind and consciousness (MAC) as the primary player, which has complete control over matter and energy (MAE).

Three possibilities are considered here:

First of course is that this is a simulated universe...that MAC on reaching a state of Superconsciousness prepares a comprehensive program with beautifully designed mathematics constants and then switches on the computer at the appropriate time and ushers in the next Big Bang...and after trillions of years...the next...and so on...a cyclic phenomena. The big obstacle to this hypothesis coming true is "the second law of thermodynamics." How do we reduce entropy from near maximum to near zero? A serious understanding of the second law of thermodynamics will reveal that if "nothing" is in control then the entropy (disorder) will either be constant or keep increasing. But, in living systems the entropy is negative, which means there is "something" in control, which causes the entropy to decrease...and "order" to increase. .

Now you might say...about "nature"...other then living systems...that entropy (disorder) is either constant or increasing... which would imply that "nothing" is in control...Here too I would say that there was "something"...such as "possibly" the Superconsciousness of the previous eon of the universe, which by a certain — one-off action — " phase transition" or something caused the entropy of the previous eon, which was obviously at its maximum to zero, which we know was the case at the beginning of the present eon of the universe.

That "one-off action" could be the "switching on" of the Big Bang by pressing a button—after preparing a comprehensive computer program—all done by that "something" such as the "Infinite mind."

You might say, "That's a tall order...How can consciousness program the universe so as to bring it back to that low level entropy, which is an essential requirement for the Big Bang to emerge?" Now, the universe is open, expanding all the time...unlikely to stop expanding and so there is no question that the entropy will not go on increasing...What shall we do to reduce entropy?

Here's the second possibility...Forget the rest of the universe...Consider just our own galaxy "the Milky Way." As the universe expands, the galaxies recede from each other...Now this may not be the case in the next few billions of years when some galaxies may indeed run into each other, and it is quite likely that Andromeda may get married to and become one with the Milky way...But in the distant future the galaxies will be so far out from each other...that you might consider each galaxy as a universe in itself, not influenced by the rest of the universe...except that it is still participating in the expansion of the universe.

Now, if each galaxy is considered as a universe in itself, its matter density is reasonably high, and the time will come to create a possible move toward a Big Crunch and thereby reverse the direction of entropy. What if the combined Superconsciousness of the entire galaxy now comes into the picture and prepares the program and designs its own "unique universe" with its own "unique laws of science?" Its own unique anthropic principle... and installs a timing device that will be switched on at the appropriate time—when the size of the galaxy is just about right so that inflation is unnecessary—to create a Big Bang and reverse the direction of entropy from order to disorder once again...from contraction to expansion...and thus create a new and "uniquely imperfect universe"...albeit a much smaller one.

It's not for me to say (remember Johnny Mathis?)...what should be the technicalities to be taken care of in regard to Big

Bang thermodynamics or the black hole/information/entropy relationships, whatever...or where the supercomputers and the control switches should be located. The Superconsciousnesses will, of course, take care of all these details.

You might ask the questions: "What if it turns out that the degree of consciousness does not reach a sufficiently advanced stage when it can reverse the direction of entropy for its galaxy, in time, before the Stelliferous Era ends?" In that case there is this third possibility, and I am most hopeful of this one:

That this omnipresent — in space and time — mind of ours can effectively come out of time and space and then travel instantaneously to far of corners of time and space of the universe. It allows the universe to proceed in its own way. But the minds can travel back in time and stay within the Stelliferous Era and have a huge choice of biochemistries from a huge variety of galaxies and times, to choose from. The degenerate and the dark eras may go on and on...but all the stars — including Hollywood ones — will be with us.

Perhaps this is already going on and that portion of the mind that is within me might have come from the sixty-fifth billion year (since the Big Bang) and it will keep traveling to and fro as it likes and when it's tired...it will go back and relax for some time in the deep future. (Note...this can also explain quantum entanglement, collapse of wave functions, observer participancy, delayed choice experiment, etc.)

In this alternative there is only one universe where lives are lived in different regions of time. The mind keeps traveling to different regions of time (within 10^{20} years after Big Bang stage) and space...entangling with biochemistries...and getting consciousness...and remains mostly in the Stelliferrous Era.

Take Your Pick

...And so in none of these possible scenarios will there be any need for us to go through those prolonged — trillions and trillions of years long — dull and useless eras mentioned earlier. Indeed... Designing a universe where life exists for less than a billionth part of its total existence is extremely uneconomical to say the least...Not at all a cost-effective design.

This entire concept outlined above has several advantages such as:

We remain most of the time in the Stelliferous Era when stars and planets are forming and life is flourishing beautifully...the only periods when there is no life is when the galaxies are collapsing and the primordial era — about 1,00,000 years — just after the Big Bang and before the Stelliferous Era...and these periods are not happening simultaneously in allgalaxies...and at any given time, there are a phenomenally large number of galaxies simultaneously alive and kicking to choose from.

We are forever moving toward an ever-increasing degree of consciousness and intelligence moving ultimately to the Superconsciousness stage — and this happens collectively as well as individually to all the minds. And of course we have already talked about the oneness of mind. It's just that consciousness always happens to be in the singular.

These ideas are predominantly neutral to all religions and have the potential to bring about the convergence of science and religion.

You might ask the question: What if I am completely wrong?

Then...As the scientist Lawrence M. Krauss says, "We have a bleak view of the universe. We started with nothing, became something, and in the end, there will be nothing again. And the nothingness is heading toward us, faster and faster everyday. And it will remain perpetual nothingness...no life, no love, no music, no science, no quantum physics)...And we learn the two lessons that we're more insignificant than we ever thought, and that the future is miserable."

It doesn't mean that Lawrence Krauss is wrong. With the current level of knowledge and with the current status of established science...most scientists would probably have similar views... The fact is that the current level of knowledge — particularly in this field, which is predominantly an uncharted territory — is at best scanty, and it doesn't help much, if scientists remain uninclined toward giving "consciousness" its due importance.

In my view, "consciousness" is important enough to be considered as a branch of physics itself...perhaps as significant as quantum physics itself.

I am forever an optimist as far as deep...very deep future is concerned...though I am not so optimistic about the immediate future...the next few centuries...where I believe we are moving steadfastly toward self-destruction...And the only way to prevent this self-destruction is to "educate" people to move toward a single religion, which is in convergence with science...And the best bet for that is a view of the future based on an understanding that there is an ever increasing degree of consciousness and intelligence, of the human mind...and that it is the same mind everywhere...omnipresent in time as well as space.

Response from a delegate...OER (6 -022): I like "scanty" as a good description of these linear conclusions about the universe within our very narrow experience—in the context of time—which completely ignores origins of dark matter and dark energy that both comprise 96 percent of our universe and which we are oblivious of.

We can't understand either in any significant detail because they both exist outside of time and our low energy state. Meaning: Dark matter and dark energy are both influenced and a result of past, present, and future reality far removed from us. And, of course, there's quantum mechanics fluctuations we can't explain either.

SK: Thanks OER...I fully agree with you...It's impossible...and in fact improper to determine the fate of the universe without a reasonably good grasp of its matter content...and the same cannot be understood without a reasonably good understanding of what is dark matter and dark energy...and these two things...as you rightly pointed out...exist completely outside the domain of "our time" and "our low energy states" as understood by us.

There are many, many other things not easily understood... for example the link between the thermodynamic arrow of time and the cosmological arrow of time. Can we indeed say with

confidence that a contracting universe would be thermodynami-
cally inconsistent? Consider, for example, our own galaxy. As
long as the internal motions inside the galaxy remain unchanged,
such as planets revolving around the sun and the solar system
revolving around the center of galaxy, what difference would it
make to our own concept of "entropy" if our galaxy instead of
moving away with the expanding universe slows down and then
reverses its direction back toward where it came from in a contracting universe.

Then there are the huge complexities involved in ascertain-
ing the relation between "information," "entropy," "black hole
thermodynamics," "the Hawking radiation," etc....not to mention the relation between positive energy of matter and motion
and negative energy of gravitational binding (we cannot be sure
that there sum is zero)...and between the properties of dark matter and those of vacuum...and we can go on and on.

Suffice to say that at this stage our knowledge is too scanty on
these matters. The best we can do is to play the game of intuition.
Let about a hundred or so scientists/cosmologists engage in a
brain storming session, following which each one of them will
write down a paragraph or two of what he or she thinks has the
greatest probability of becoming true, and then somebody can
compile the views and arrive at some sort of consensus.

If I were one of these, this would be my paragraph:

"That this universe of ours is a designed universe, designed
with constants so finely tuned so as to enable life to evolve and
theoretical physicists to emerge and understand the universe.
And if it is a designed universe, it must have been designed
by consciousness/Superconsciousness, whatever...There is no
way it could be designed by unconsciousness. And if it is a designed universe to accommodate life, it is impossible for life to
exist for just a trillionth part of the life of the universe...that
would be a highly uneconomical design...and there must be
some way to ensure that either those prolonged eras (the degenerate era, the black hole era, and the dark era) are somehow
cut short... bydark matter/dark energy or something similarly
exotic...or the mind and consciousness can keep traveling back

and forth between the Stelliferous Era and the above dull eras to ensure life to exist on a continuous basis…And most important of all is my belief that the degree of intelligence and consciousness is forever increasing."

CONCLUDING THOUGHT

There is "something" that controls nature such as an all-intelligent "infinite mind." We don't have to call it "someone.," but we can call it "God." The concept of an "infinite mind" would have been accepted by Einstein if we recall his extremely thought provoking talk of the day before on "cosmic religious feeling."

Similarly, it would have been surely accepted by Erwin Schrödinger, if we recall his talk on the "Oneness of Mind...the Arithmetical Paradox" In fact, this hypothesis would be acceptable to all who have spoken during the seminar until now.

This concept of an "infinite mind" makes "immortality" (not of "this life" but of a "life in general" kind) a possible hypothesis. In my view: "It is impossible not to be aware that we are aware... or to be aware that we are not aware." If there are intervening times...in between successive consciousness states...we are unconscious of these unconscious tenures...and they pass quickly. This concept, along with the "six words," is the trump card that can unite all religions into one and pave the way to end all conflicts that may be attributed to religious extremism.

However, if we take the laws of causation seriously enough... the realization will manifest itself that no war or conflict can actually be attributed to religion...for the simple reason that no religion teaches war...thus it can be said that all those who engage in conflicts by considering their religion to be superior to others cannot be called religious at all.

With all this, a scientist—however knowledgeable he may be—cannot put himself on a pedestal as if he has a complete understanding of science and hence represents "science." Likewise he cannot consider people involved in conflicts (attributed to religious extremism) as representing "religion" and then put them on the floor…and talk about religion…and tell the world that religion is at the root of "evil," without understanding the essence of "religion."

The primary essence of "religion" is the belief that there is something that controls nature…and this understanding and belief provides solace and peace of mind via prayers and meditations.

Now there are those who believe that this something is actually "someone" such as a superhuman being and they call him "God." While this is an impossibility, no harm is done if this understanding gives solace and peace of mind to them…provided they do not consider their religion to be superior to other religions and are intolerant to them, in which case they cannot be called religious at all…as no religion teaches intolerance.

And then there are those who believe neither someone nor something is in control and that nature acts on its own. Again, no harm is done if they do not need a controller and are always at peace with the world with this understanding.

All three categories can live with each other and with the environment in harmony…and call themselves "religious." In my opinion, the first category of "something" is more likely to be the controller rather than "someone" or "nothing."

No scientist can explain "reality" without an adequate understanding of the role of consciousness…it's absolutely futile… There is a substantial limit to what the scientists can achieve by talking only of matter and energy, and particles and fields, and ignoring consciousness.

An invisible, nonphysical, intelligent force is responsible in creating consciousness after interactions with an appropriate biochemistry and then tying up the "conscious" mind with a contraption called brain.

The percentage of population that believes in a physical-function God or Gods. though still substantial, is reducing as people begin to understand this intelligent force. The notion that God is a mind (Paul Davies)…an all intelligent infinite mind…is a trump card notion…that has the potential to bring about a convergence of all religions into one (for who can object to it?)…and end all religion-based conflicts.

Consider the statement:

"It is impossible for me (SK) to be aware that I am not aware or to be not aware that I am aware. There may be intervening periods of unconsciousness...but I will not be conscious of these unconscious tenures, which means they will pass quickly...rather instantaneously...and I'll be conscious again except that I'll have a different body with a different name." If the above statement is not true then I who am in my seventies may live a few more years and then...There is nothing any more...No life...No existence...No universe...That's it...And if this is the future for me (SK), it is the same for you and everyone else...It is the same for Einstein...for Newton... for Schrödinger...for Kennedy...not to mention Jesus and Gautama Buddha, etc. Perpetual nothingness is in store for all of us...

A little analysis will show that it is impossible for the statement "A" to be untrue...and this is well taken care of by the concept of an "infinite mind" and the "six words."

An infinite mind "in control" is less difficult to believe than a "complex yet organized infinite universe" materializing on its own as if by magic...The all-intelligent infinite mind does not create life by magic...rather it creates the mechanism of evolution that creates life.

The dictionaries of the world must redefine the meaning of God...The words "superhuman being" may be replaced by the words "intelligent infinite mind." The atheists can be happy... for a physical-function God such as a "superhuman being that controls nature" does not exist. The theists can be happy for there is definitely something such as an "infinite mind" that controls nature...and they can call it "God."

The atheists can shake hands with the theists...It's so simple. Where is the need for a conflict? Essence of religion lies in living and letting live...in love and compassion...in assuaging pain of others and taking care of the environment...in obtaining solace and peace of mind via prayers and meditations...In praying to your God...What if he is invisible?

But...the infinite mind plays dice...it controls nature but has no control on "probability"...it will not come to our rescue if we are moving toward self-destruction. We must create appropriate probabilities to prevent the human civilization from collapse.

That's why this seminar...

**

Lunch Break

**

ENDNOTES (6-016 TO 6-022)

6-016:

Quotations/statements shown at the beginning of chapter six E...i.e., "six words." The last session."

6-017:

The six words can be decoded from the quotations/statements above.

6-018:

The discussion that ensued after the tea break is something that actually took place between the commenters "Charliefoxtrot999," "Diogenes of Alaska," and "SKSagar" in response to a blog by the scientist Mario Livio entitled "Can Black Holes Tell Us Something about Digital Computers?" (Posted: 01/08/2013 5:04 p.m.)

To read the blog and the responses refer "www.huffington-post.com/Mario Livio/Can black holes tell us something about digital computers."

6-019:

The actual word used was different.

6-020:

The actual words used were different.

6 -020a :

Ref..``http://www.huffingtonpost.co.uk/2012/10/11/physicists-may-have-evide_n_1957777.html?just_reloaded=1``

6-021:

For further reading, refer to the responses to a blog by the scientist Louis A. Del Monte entitled "Dark Energy Explained — A New Theory." Posted: 03/11/2013 2:12 p.m. on huffingtonpost and the discussion between the blogger and the commenter "SKSagar." To read the blog and the responses refer to "www.huffingtonpost.com/LouisADelMonte/DarkenergyexplainedANewTheory/SKSagar."

6-022:

For further reading refer to the responses to a blog/video by Cara Santa Maria entitled: "Lawrence Krauss 'A universe From Nothing'" (Posted: 07/18/2012 7:53 a.m. Updated: 07/18/2012 8:07 a.m.) and the discussion between the commenters "SKSagar" and "Oneeasyrider."

To read the blog and the responses refer to "www.huffingtonpost.com/CaraSantaMaria/LawrenceMKrauss/Auniversefromnothing/SKSagar."

**

Chapter Six D

THIS CIVILIZATION MAY COLLAPSE...UNLESS

The Post Lunch Session

The hoarseness in the voice had reached uncomfortable levels...
Yours truly took steroids during lunch break and returned on the
stage after a half hour of complete silence...Absolute concentration
was required to ensure that that something remained in the head
all the time, supplying it with the flow of information/thoughts
in a continuous way. The ensuing session was a difficult one.

The direction of the topic needed to be reoriented to stress
on the importance that the realization must manifest itself that
human civilization is on the verge of collapse unless appropriate
probabilities are created to prevent such a collapse.

Convergence of religion with science...and consequently the
convergence of all religions into one religion is a fundamental
requirement to ensure that these probabilities remain at a reasonably high level. The concept of an "infinite mind" and the "six
words" are the best bets for the same.

This Civilization May Collapse...Unless:

"If 'intelligence' does not overpower "the ego" within the next fifty years, then the Technological Civilization (TC) of the planet Earth is destined for self-destruction within the next five hundred years." Human civilization is under tensile stress and cracking fast. Consider this as a technical problem. Intelligent minds of the world must come forward and reduce the stress, seal the cracks, and prevent the collapse.

The problem is hugely complex and beyond the capabilities of politicians and world leaders to resolve, without scientific analysis and an unbiased mind. No doubt there are innumerable experts scattered worldwide, who write "eye opening" articles to facilitate a better understanding of the affairs of the world, particularly with regard to the conflicts between nations. But the combined expertise and resourcefulness of the politicians and world leaders who matter, including that of the experts, is not enough to generate solutions to end these conflicts or even reduce their magnitude. This is borne out by the fact that appropriate solutions are not forthcoming. One might ask the question — how does one evaluate in mathematical terms, whether there is progress in this regard? What is the current status of conflicts on the planet and what are the probabilities of collapse of human civilization in a fixed period of time, say within the next 250 years? What parameters should be selected in the mathematical formulations to find answers to the questions given above? Should there be a committee of specialized experts to keep track and determine at regular intervals the probability index of a nuclear war taking place anywhere in the world in a fixed period of time, say in a calendar year, and for each nuclear attack the probability index of a chain reaction leading to a global war? Is the rising population a factor that may indirectly lead toward increased probabilities of nuclear wars? An attempt will be made later...to understand these issues. But understanding the problems is only one part, arriving at solutions, which are acceptable to all concerned, by conquering egos with intelligence is the other and more difficult part. It is here that a realization must manifest itself that only the highest quality minds on the planet earth have a chance to resolve the problems of the planet. At present most of the top analytical minds in the world are scientists, who are engaged in the pursuit of science and not engaged in this "redesigning and retrofitting of the structure." These top brains of the world must change their priorities now. It will do no harm to the planet if they forget science for some time and devote all their time and resources to this "redesigning the structure." The collapse of civilization must be avoided at any cost. Of what use is relativity theory or quantum mechanics if there will be no one left alive?

There is only one way that human civilization can be prevented from self-destruction, and that is "urgent reversal of the rising trend of world population." The realization to this effect may not manifest in time before one or more nuclear wars wipes off large chunks of population, particularly from the densely populated zones. Some important and responsible nations of the West may be aware of this likelihood and may present a blind eye to such developments that may lead toward a nuclear war between nations in conflict, particularly so if these conflicting nations are significant contributors to the population explosion.

A study of the nonuniform growth of populations in different countries will reveal that if the same trend continues, the combined population of India/Pakistan, which is currently about four times the combined population of the United States/Canada will be eight times one hundred years hence, seventy-five times four hundred years hence, and about thirty-five hundred times seven hundred years from now.

Another exercise will show that if the same trend continues, then the five countries Nigeria, Bangladesh, India, Sudan, and Pakistan will leave all other countries way behind and together will constitute 99 percent of the total population while occupying only about 10 percent of the total area, whereas the remaining 1 percent will occupy 90 percent of the area.

(This looks impossible, but its true. The 2009 Penguin year Book referred for this.For the abovementioned five countries the total population in 2008 stood at about 1.65billion and the weighted average annual growth rate of population was 1.71%, so if the same trend continues the population would in 700 years, reach 240000 billion. For the rest of the World the total population in 2008 was 5 billion and the weighted average annual growth rate of about 0.85 %, the population in 700 years would grow to about 2000 billion.The corresponding figures for China are : Population in 2008.. 1.33 billion, annual growth rate 0.68%, projected population in 700 years 108 billion, and for the USA : Population in 2008 ..0.304 billion, annual growth rate 0.88%, projected population in 700 years 140 billion.)

If no steps are taken to reverse the trend of rising population, then a simple calculation will show that in some areas of the world the density of population just five hundred years from now may be as mind boggling as twenty people per square meter, which means the entire land area filled with people literally touching each other.

Surely this cannot possibly happen. Nuclear attacks will become imminent long before that stage is reached. The notion that large-scale conventional wars may help in reducing population levels is not correct. The twentieth century witnessed several wars including two World Wars and the number of lives lost was unmatched in any previous century, yet the real population explosion took place in the twentieth century. In fact, the rate of increase of population experienced a sudden surge around the year 1930. Nuclear wars followed by chain reactions are the only way of causing sudden massive falls in world population, but it is also the only way of causing the extinction of the human race from the planet (apart of course from cosmological causes such as a major asteroid hit).

Even assuming that the probability of a nuclear attack during a year is as low as 1 in 200 (assumption "a1"), i.e., 0.5 percent and rises each year (due to increase in population) by as little as 0.01 percent in the first two hundred fifty years and by 0.02 percent per year thereafter and assuming further a 25 percent probability of a global chain reaction (Superpower confrontation), for each nuclear attack (assumption "a2"), then the probable chance of the TC self-destructing itself within the next 250 years is worked out to be in excess of 50 percent, as shown:

PERIOD	PROBABILITY OF A NUCLEAR ATTACK	PROBABILITY OF EXTINCTION OF HUMAN RACE
BETWEEN 2014 AND 2064	(1 - 0.995 X 0.9949 X 0.9948 X ----------X 0.9901) 100 = 31.4 %	(0.314 X 0.25) 100 = 7.85 %
BETWEEN 2064 AND 2114	(1 - 0.99 X 0.9899 X 0.9898 X---------X O.9851) 100 = 46.7 %	(0.467 X 0.25) 100 = 11.7 %
BETWEEN 2014 AND 2114	(1 - 0.686 X 0.533) 100 = 63.4 %	(1 - 0.9215 X 0.883) 100 = 18.6 %
BETWEEN 2114 AND 2164	(1 - 0.985 X 0.9849 X 0.9848 X ---------X 0.9801) 100 = 58.6 %	(0.586 X 0.25) 100 = 14.7 %
BETWEEN 2014 AND 2164	(1 - 0.686 X 0.533 X 0.414)100 = 84.8 %	(1 - 0.814 X 0.853) 100 = 30.6 %
BETWEEN 2164 AND 2214	(1 - 0.98 X 0.9799 X 0.9798 X --------- X 0.9751) 100 = 68.0 %	(0.68 X 0.25) 100 = 17.0 %
BETWEEN 2164 AND 2214	(1 - 0.152 X 0.32) 100 = 95.1 %	(1 - 0.694 X 0.83) 100 = 42.4 %
BETWEEN 2214 AND 2264	(1 - 0.975 X 0.9749 X 0.9748 X --------- X 0.9701) 100 = 75.2 %	(0.752 X 0.25) 100 = 18.9 %
BETWEEN 2014 AND 2264	(1 - 0.049 X 0.248) 100 = 98.8 %	(1 - 0.576 X 0.811) = 53.3 %

Chilling thought.

It goes without saying that this rising trend of population will not be sustained and something is bound to happen to reverse the trend. There are only two possibilities that come to mind:

Good sense will prevail, countries of the world will come together to formulate and implement plans to take adequate measures to reverse the trend with incentives for good compliance and sharp disincentives for noncompliance. Before that, conflicts and terrorism must cease. "P1"

Nuclear Wars — conventional wars will not do. There were two World Wars and several other wars between nations during the last century, but the sharpest population increase was also recorded in the last century. "P2"

The former (P1) is not happening; there is no sign of it, and no one is talking about it. The latter (P2) I believe is on course, indirectly, of course.

From the above...considering the second assumption (a2)... the probabilities of self-destruction of the human race are worked out as: 25 percent after the first attack; 44 percent (cumulative probability) after the second attack; 58 percent (cp) after the third attack; 69 percent (cp) after the fourth attack and so on.

Chilling thought.

It goes without saying that, we have no choice whatsoever, other than bringing down the probability index...as per the first assumption "a1" from one in two hundred (per year) to as close to as nil as possible.

What is the current status?

The current geopolitical situation in the world is such that this assumption of an approximate one in two hundred chance of a nuclear war taking place in a calendar year may not be an incorrect assumption, and if it is indeed somewhat incorrect it may need an upward revision rather than a downward revision. These developments including the blind-eying could be an indirect plan, designed to force the realization that there are no options other than a collective approach by the leaders of the world's nations to come together and address the problem of population explosion. It is profoundly imminent that at some time in the future this realization will manifest itself. It remains to be seen if that happens before the first nuclear war, or after the first, second, or third such war. Nations in conflict should become aware of these possibilities and understand the consequences of not resolving their disputes in quick time. Leaders of these conflicting nations must come together with a positive attitude and a neutral perspective and address their problems in the same way as a neutral person, such as a being from outer space, would have done.

How would such a neutral figure (let's call him "F") who has watched all the proceedings and has a complete knowledge and understanding of all the conflicts on the planet Earth address the problems in a way that is as far as possible acceptable to all concerned? It is assumed, of course that "F" understands human nature and is perfectly aware that a typical human being anywhere in the world wants to live in peace and that the average moral standard is the same in all countries and does not differ from country to country as it is not a measure of "temperature" or "humidity" The average quotient "qx" pertaining to the desire to "live and let live" is also generally the same for all countries provided the corresponding quotient "qy" pertaining to the leaders of the country is greater than or at least equal to "qx." Our man understands science and knows that while laws of probability govern what happens at the quantum level, at the level of tables and chairs, animals, you and me, it's the laws of causation that govern what happens, and it is this phenomena of cause and effect and the interactions of the world, which lead to the coming together of millions of atoms and molecules becoming a speck of jelly. It is the same phenomena that is responsible for the said speck of jelly to go on to become either Hitler, or Gandhi, or Einstein, or you and me. Each of us is subject to a specific and unique set of interactions, memories of which keep getting stored in our brains, and all our actions are dictated by these interactions. Carrying forward this logic, our man therefore understands that a person with a substantially small "qx" value (a murderer or a terrorist for example) is not responsible for his actions, but from practical considerations it is advisable not to have tea with him (Einstein's words) The only way out is by raising his "intelligence" to a level that exceeds his "ego" by remote control means through a proper understanding of his past interactions, which caused his "qx" to fall. In view of the above, while the word "terrorism" may still be used, the word "terrorist" should henceforth be replaced by a set of three words: "Misunderstood Human Being"(MHB), as these terrorists happen to be just victims of time.

I ask the question again…What is the current status?

There is as yet no sign of P1 taking place; there are at least four regions in the planet where there is a distinct possibility of a nuclear war originating, such as Pakistan, North Korea, Iran, and Israel. This must be avoided at all costs…by eliminating conflicts between nations and between religions. One by one, the problems faced by each of the above nations…and consequently by those nations, which are in conflict with them, must be understood, analyzed, and resolved…

The realization must manifest itself that the geopolitics of the world as a whole must be understood properly, and each and every problem must be addressed absolutely from a neutral perspective.

With all this it does not serve the purpose if we call a certain country — which is just a geographical area — as a terrorist state and another country — another geographical area — as a "selfish and arrogant policeman," when we know that it is the leaders of nations who must be held responsible for bringing their nations to such disrepute. Therefore "F" should first understand how the common man from the various countries finds himself entangled in these complex and dangerous situations for no fault of his and how he can be extricated from these entangled states.

Now, there is no such being from outer space "F." To understand the problems of the planet and suggest solutions and the way forward. What is the next best alternative? Arriving at this alternative is primarily the objective of this seminar, and the role of all the beautiful minds in this seminar is to arrive at solutions precisely in line with expectations of what "F" would have done.

The solution to the problem must be aimed at a package of measures, which, if carried successfully, will lead toward a world conference or an Earth Summit to be attended by responsible world leaders from all countries with representation having weight/age in accordance with countries/groups of people most affected by conflicts. This must be organized. All attendees no matter which country or religion or group they represent shall have an unbiased and accommodating approach with "securing peace for all" and "globalization" as the ultimate objective. The Earth Summit shall be preceded by papers written by some of the world's leading scholars and scientists including some Nobel Laureates in peace, having in-depth knowledge of past history with an unbiased and neutral approach as expected of "F." The Earth Summit should comprehensively outline action to be taken by all concerned. It should be an annual affair with a built-in mechanism to monitor, review, tackle noncompliance effectively, and ensure the timely fulfillment of all the objectives identified.

This seminar is primarily aimed at an urgent realization that an international conference of the kind envisaged above must be organized by responsible leaders of the world, where the wise people can explain to the misguided ones what they should do for their own good and for the good of the world, and the world leaders should work to create an environment where the misguided people can do what the wise men have asked them to do. In short, a conference aimed at allowing intelligence of the planet, which has been trailing badly, to start lengthening its strides and recover lost ground quickly so as to overtake the *ego* before the latter reaches its winning post of destruction. In the first few seminars it may not be practical for the misguided ones to attend (We understand them well but we cannot take tea with them yet). This should not pose any problem; they will be watching all the proceedings live on television wherever they are. Later on when their I/E ratio exceeds unity they can be full-fledged participants and, who knows, they may know a few things that the wise men don't. What if the conferences don't happen? What if nothing else happens to control the world and the conflicts continue and the rising trend of population continues? What if this highly advanced technological civilization with its brilliant minds, its scholars, philosophers, quantum physicists, cosmologists, Nobel Laureates, mathematicians, statisticians, IT experts, programmers, politicians, and great leaders, etc., failed to diagnose the problems and prescribe a suitable treatment for survival?

Well...three hundred years more or less is all we can hope for, which means we can come back and live for just a few more lives — perhaps a maximum of four — and then it's all over — no more love, happiness, memories, no more music, cricket or movies, no more quantum physics, no more cosmology, no more understanding the universe — just endless waiting for millions of years for another perfect biochemistry, and perfect conditions for evolution of human life materializing again, and then wait for hundreds of thousands of years more to reach the current level of intelligence and then hope that "intelligence" will have at least a slender lead over "ego."

Tea Break

**

During all this time, the three ladies who had earlier come prepared with a draft manuscript of their presentations were taking stock of the situation and kept modifying their manuscripts in the light of the proceedings of the seminar.

In the remaining part of the third day, the twenty-five distinguished speakers assembled in a separate conference hall at 9 a.m. The three ladies then gave their presentations to this distinguished gathering on each of the topics given below:

Formation of a Global Organization Structure to Control Life on the Planet Earth.

Resolution of Conflicts of the World.

Control of World Population.

The presentations included a comprehensive package of measures/actions needed to be taken by the leaders of nations in conflict as well as by the "global organization" to ensure such implementations. The package of measures also included "how to deal with misunderstood human beings who were once normal human beings who became victims of the times created by the conflicts" as well as how to deal with the "nuclear arsenal of the world."

At the end of each presentation, there was a discussion and an exchange of views. Thereafter, the presenters incorporated these views in their paper, by exercising their own judgment and to the best of their abilities completed the document. This took up the entire remaining part of the third day. During that period the delegates in the main hall were entertained with a cultural program, appropriately designed with a global perspective aimed at uniting the religions of the world.

In the pre-lunch session of the fourth day, the three documents were read out in the main hall. Each presentation was followed by a question-and-answer session followed by feedback from all the delegates in a prescribed format. Thereafter, the twenty-five distinguished speakers assembled again in the conference room. They were joined by a select group of delegates — from those who either gave positive and useful suggestions in the feedback or asked pertinent questions elucidating gainful responses from the presenters — from the audience. The feedback documents were scrutinized on a statistical basis and each paragraph of the main document (each one of the measures included in the package) was reviewed and either retained as such, or deleted or suitably modified, and the document frozen. This exercise was carried out for all the three documents referred above and a composite document compiled.

Copies of the compiled document, which included in comprehensive detail all that was spoken on the first three days plus the three documents referred above, were distributed to all the delegates present. This document constituted the proceedings of the seminar.

The seminar was entitled "Convergence of Science and Religion to Escape Self-Destruction."

**

And then came the final session...a general-purpose question-and-answer.

**

Chapter Six E

SIX WORDS—THE LAST SESSION

After lunch on the last day, the delegates started returning to their seats. The twenty-five who were in the dais took their appointed seats in the front row of the auditorium. Classical music played in the background...The giant screen in front displayed images of spellbinding beauty...Each image was displayed for about a minute and then gradually disappeared in about fifteen seconds and the next one appeared...Against each image was a caption...The number of words in each caption corresponded to a letter in the English alphabet: (For example 1 =a,2=b, etc)

**

"My body functions as a pure mechanism according to the laws of nature, yet I am the person who controls all the motions". (Erwin Schrödinger)

It's all in the numbers.

**

Note

"The personal self equals the omnipresent eternal self" is the grandest thought". (Erwin Schrödinger)

The concept of the omnipresent mind leads us toward a cosmic consciousness.

***

All religions are created by laws of science.

And

In the eyes of the quantum-level God as an observer, there exists no conflict between religions or between science and religion.

Classical interactions create quantum probabilities

****

"In this excessively enlarged body, the spirit remains what it was, too small to fill it, too feeble to direct

This increased body awaits a supplement of soul

The mechanism demands a mysticism." (Louis de Broglie)

***

We are part of a technologically advanced civilization; we can convert potential energies into motion, and we are aware

That

One little gram of uranium gives us more than ten tons of coal

That precisely is the problem.

****

"Humanity groans half-crushed under the weight of the advances that it has made

It does not know it makes its own future.

It is for it to make up its mind if it wishes to live." (Louis de Broglie)

Intelligence must conquer ego.

****

ENTER THE "MOC" AND THREE SCIENTISTS OF TODAY (NOT THE REVISED VERSIONS) TO ANSWER QUESTIONS

Q: Is the anthropic principle Earthcentric? Or in other words... Is human consciousness something special...Are we (late-comers on the scene) indeed special and will we survive from self-destruction?

MOC: Of course we are *not* special...we *cannot* be special...The universe is too large for that possibility. It's like considering a single ant as the only inhabitant on the planet Earth...Not at all a cost-effective design.

But the realization having manifested itself that we are not special could imply that in some ways we are indeed special for the simple reason that such a realization that we are not special has manifested itself...And the fact that *now* the said realization has manifested itself that we are indeed in some ways special or at least among those who are special, leads us to another realization...That we are not "so special" that a divine universal force will arrive to save us from extinction...but "special enough" to have been granted the realization (perhaps by the divine universal force)...That it is up to us to create certain probabilities on the planet Earth that can enable us to survive from self-destruction.

It's the same thing with "free will." In essence, we do not have free will...for it's the interactions of the world on us that make us what we are. But the realization has manifested itself on us — due to these interactions — that we do not have free will and implies that *now* we have the free will to do what we like to do...And maybe improve our future interactions.

Q: What exactly is the relationship between the human consciousness and the physical world?

Roger Penrose:

"I believe that consciousness is a physically accessible concept. It seems to me that there are at least two different aspects to consciousness. On the one hand there are passive manifestations to consciousness, which involve awareness, and on the other hand there are its active manifestations, which involve concepts like "free will" and the carrying out of actions under our free will. And then there is something else...perhaps something in between...I refer to the use of the term "understanding" or perhaps "insight," which is often a better word...And then of course there are the two words "awareness" and "intelligence." It seems to me that "intelligence" is something that requires "understanding" and "understanding" is something that requires "awareness" and it seems to me that...Appropriate physical action of the brain evokes awareness but that this physical action cannot be properly simulated "computationally." And, according to one aspect of this last assumption, it seems there is nothing we need look for outside known physics in order to find the appropriate noncomputational action...and according to another aspect of the same assumption our physical understanding is inappropriate for the description of "awareness." This is something outside known physics. Maybe future science will explain the nature of consciousness. But present-day science does not" (6-023)

Q: Does mathematics have a role to play in the manifestation of "reality?"

RP:

Mathematics is at the heart of science, which is at the heart of understanding...It is an integral and the most important part of science...So much so that if we go deeper and deeper and probe deep down into the laws of nature, we will discover that the physical world almost evaporates and just about only "mathematics" remains...

Q: But are not mathematical symbols merely tools to describe what we really mean?

MOC: It doesn't look like that...it looks more like mathematics dictates terms to physics.

Consider the law of universal gravitation...there are bodies "A" and "B" having masses "Ma" and "Mb" separated by a distance "d" exerting forces on each other...and there is "g," a universal constant...and there is an equation linking the force "F" with these parameters:

$$F = g * Ma * Mb / d^2$$

No doubt, the law was determined by physicists—who knew mathematics—by observation and experiment and logical thinking and analysis, whatever...But it is a Mathematics law...it's mathematics all the way through and through and it is telling physics what to do.

Indeed..."mathematics" is the President and the company is called "The universe."

And, of course, "mathematics"—comprising mathematics objects—is an aspect of reality, which cannot be ignored when we are talking of reality, as much as we cannot ignore "matter"... or "mind."...or "consciousness" or "information," etc.

Who knows when—in the deep future—when dark energy drives the universe into absolute nothingness...what will that energy get converted into?...perhaps not into void...

What do we know?

But the President knows.

Q: Should intelligent design be taught alongside Darwinian evolution in schools?

Charles Townes:

People are misusing the term "intelligent design" to think that everything is frozen by that one act of creation, and that there is no evolution, no changes. It's totally illogical in my view. Intelligent design, as one sees it from a scientific point of view, seems to be quite real. This is a very special universe. It's remarkable that it came out just this way. If the laws of physics were'nt just the way they are, we couldn't be here at all. The sun couldn't be there; the laws of gravity and nuclear laws and magnetic theory, quantum mechanics, and so on have to be just the way they are for us to be here. (6-024)

Q: Is it not possible...that the constants aren't tuned to the universe...that the universe, which evolved on top of the constants, is tuned to them...That the universe isn't tuned to us, we, who evolved in this universe, are finely tuned to the universe...That the universe evolved on top of the constants, and is tuned to them...that it means that the constants were supplied earlier and the universe evolved on top of them?

Is it not possible...that the "carbon-based life," as we are used to, is not a specific requirement for consciousness to evolve...or maybe the laws of science and even the elements are completely different in different universes...a measure of probability?

Is it not possible...that even the second law of thermodynamics is also not applicable in some universes?

Is it not possible...that matter and energy (MAE) is the primary player and responsible for the occasional appearances of mind and consciousness (MAC) at various locations and times, whenever — and wherever — it has "randomly" created a certain biochemistry (whether carbon-based or not) for MAC to arrive?

Is it not possible...that randomness is supreme and the kind of life that is commensurate with the constants provided (also chosen randomly) materializes?

Charles Townes:

Some scientists argue that "Well, there are an enormous number of universes and each one is a little different. This one just happened to turn out right." Well that's a postulate, and it's a pretty fantastic postulate—it assumes there really are an enormous number of universes and that the laws could be different for each one of them. The other possibility is that ours was planned, and that's why it has come out so specially. Now that design could include evolution perfectly well. It's very clear that there is evolution, and it's important. Evolution is here, and intelligent design is here, and they are both consistent. (6-024)

MOC:

These alternatives (where MAE is the primary player) as assumed in the questions do not explain how and from where does "intelligence" arrive on the scene. Perhaps it is an extremely rare commodity and also perhaps the quantum of intelligence is extremely limited...certainly not enough to prevent itself from self-destruction, and definitely not enough to avoid getting obliterated by a major asteroid hit...And if indeed...the universe is not a designed one...And if MAE is the primary player it would mean there are millions of universes to cater for the enormous range in which the constants varied...but only about one in several million universes had the desired constants to enable life and consciousness to evolve...the remaining exist without anyone being conscious or aware of it, let alone "collapse their wave functions."

In such cases...About one in several million universes supporting "life and consciousness"...and even in that one... "Life and Consciousness" lasting out for just about one unit of time out of nearly 10^{80} units of time...and as if that's not enough...even in that unit of time we are going all out to self-destruct ourselves; I think it is a staggeringly pessimistic viewpoint.

Q: If science and religion share a common purpose, why have their proponents tended to be at loggerheads throughout history?

Charles Townes:

Science and religion have had a long interaction; some of it has been good and some hasn't. But science has been digging deeper and deeper, and as it has done so, particularly in the basic sciences like physics and astronomy, we have begun to understand more. We have found that the world is not deterministic; quantum mechanics has revolutionized physics by showing that things are not completely predictable. That doesn't mean that we have found just where God comes in, but we know now that things are not as predictable as we thought, and that there are things we don't understand. For example we don't know what some 95 percent of the matter in the universe is, we can't see it, it is neither atom nor molecule apparently. We think we can prove its there, we see its effect on gravity, but we don't know what and where it is, other than broadly scattered around the universe. And that's very strange.

So as science encounters mysteries, it is starting to recognize its limitations and become somewhat more open. There are still scientists who differ strongly with religion and vice versa. But I think people are being more open—minded in recognizing the limitations in our frame of understanding. (6-024)

Q: The analysis...by "SK" in his talk...that there is a 50-percent probability of self-destruction within the next two hundred fifty years...how far is it reliable? Are the values of the two parameters chosen— 1) A one in two hundred chance of a nuclear attack per calendar year and 2) A one in four chance of a global chain reaction (with superpowers entering the fray, and confronting each other) taking place for each nuclear attack, realistic or are they exaggerated—too pessimistic?

MOC: Yes...this is a pertinent question...There can be conflicting and widely varying views on the subject...I guess we'll let James Martin answer this question.

James Martin:

"The twenty-first century is confronted with one of the greatest paradigm shifts in history — *There will be no all-out war between nuclear nations or no civilization.*

Henry Kissinger observed that the greatest danger of nuclear war lies not in the deliberate actions of wicked men but in the inability of harassed men to manage events that have run away from them. This surely describes the future.

The world came shockingly close to nuclear war in 1962 with the Cuban missile crisis. In the defining moments of the crisis, the United States blockaded a Soviet fleet to prevent it from going to Cuba. The executive committee of the United States who called the shots was unaware that the four Soviet submarines had nuclear weapons. At about 5 p.m. on October 27, 1962, an American ship, depth charged a Soviet submarine, unaware that it had a nuclear weapon on board. The depth charge exploded next to the hull but didn't penetrate the hull. The Russian captain felt honor bound to retaliate and ordered that a nuclear weapon be launched at the Americans. To do so, two other officers had to agree to the firing and turn their keys simultaneously. At the last moment, the Second Captain Vassili Alexandrovich Arkhipov refused. If he had not done so there would have been a devastating nuclear war.

In the 1980s, the game sped up drastically. Cruise missiles — designed to carry nuclear warheads immensely more powerful than the Hiroshima bomb — were built. The nuclear command and control systems were designed so that at the very highest level of alert, nuclear retaliation would happen automatically with preprogrammed missiles. The situation becomes like that of the gun fighters of the classic westerns. Each wants to be "the fastest gun in the West" instantly ready to fire.

A major problem with the control of nuclear forces, however, is that the brains would be the first thing to be attacked. What happens if Washington and the US president are destroyed?

What happens if the command and control systems are destroyed? When the top-level commanders are destroyed, lower level commanders have to take over, but then how are unauthorized firings prevented? (6-025)

Biological warfare—after a nuclear attack—with the release of large number of small bomblets spraying smallpox, anthrax, and deadly plagues on a massive scale to wipe out any shattered survivors of the nuclear devastation, is a distinct possibility.

A few hundred nuclear warheads are enough to destroy the entire human civilization and we have accumulated nearly a hundred thousand of them on the planet. And there is no moral force acting against this…No outrage shown by the Church…or by religious leaders of any religion against this massive accumulation.

Nations in conflict are always in a state of readiness…and if a suspicious activity is detected…a preemptive strike to prevent from being attacked becomes highly likely.

With all this…Perhaps a one in two hundred chance of a nuclear attack—anywhere in the world—in a calendar year looks somewhat incorrect—on the wrong side—perhaps one in one hundred might be more accurate…And even if we consider a reduced likelihood, say one in six (This parameter value is extremely difficult to predict) of a global chain reaction taking place after each nuclear attack, we may not survive to see the.twenty-fifth century

Q: How do we reduce these indices?

MOC: The conflicts between nations must be resolved with immediate effect and the resolutions must be based on neutral perspectives.

Q: How do we do that…can you elaborate with examples?

MOC: There are plenty of them…Refer to the "Document."

Q: Are the drone strikes in Pakistan acceptable for peace for all concerned?

MOC: I guess I'll ask SK to answer this.

SK: As I said earlier: The problem must be understood from a neutral perspective:

How would a neutral figure such as a being from outer space (let's call him "F") who has watched all the proceedings and has a complete knowledge and understanding of all the conflicts on the planet Earth address the problems in a way that is as far as possible acceptable to all concerned. With all this it does not serve the purpose if we call a certain country—which is just a geographical area—as a terrorist state and another country—another geographical area—as a "selfish and arrogant policeman," when we know that it is the leaders of nations who must be held responsible for bringing their nations to such disrepute.

Therefore "F" should first understand how the common man from the various countries finds himself entangled in these complex and dangerous situations for no fault of his and how he can be extricated from these entangled states.

Where did the trouble start?

Consider the United States of America.

Once it was considered the best place in the world to be in.

In the words of Einstein: "it has an international 'psyche,' it constitutes the bulwark of the democratic way of life, it has demonstrated that individual freedom provides a better basis for productive labor than any form of tyranny, its political and economic position is so powerful that it can help the world by breaking the tradition of war from which the world suffered."

All that has changed now. Having become rich and powerful, America also tended to become selfish and arrogant, selfish in the sense that it started grabbing a disproportionately higher share of the world's resources and arrogant in the sense that in order to ensure the safety and security of its own people, America took recourse to large-scale interferences in the affairs of other countries.

It is no secret that the American leadership—of a certain category—made common cause with dictatorships of several countries of the Islamic world in capturing resources for itself (cheaply), with little or no regard for the common people in these countries; it allowed the dictatorships to continue and delayed the formation of democracies for as long as it could...and now that the trend is changing with the common people revolting against the dictators, it is changing sides and making common cause with the revolutionists.

Now consider the predicament of the common man in Pakistan. He was doing fine for several decades after getting independence — never mind sometimes fighting and sometimes playing cricket with his Indian friend — but suddenly found terrorists in his land...toward his west...Where did they come from? Driven out from Iraq and Afghanistan of course...And then to compound it all, a certain section of the population with a different (and dangerous) ideology decided to support the terrorists... not to mention play host to them.

What happens — when a drone strikes?

Its all in line with expectations...It kills people — intentionally designed to be killed — not because there exists a remote possibility that they might have been involved in the 9/11 attacks...but because of a remote probability that they may in future make plans for terrorist strikes. It kills innocent people — not intentionally designed to be killed — as the drones cannot distinguish to which category the victims belong to.

It puts the common man in Pakistan exposed to an ever-increasing probability of domestic terrorism. It reduces the already depleted chance of a reconciliation with the Taliban and widens the rift between the two factions of Muslims.

It puts the leaders of Pakistan in a hopeless situation...How will they face the public?...They may pretend to condemn the strike and show that it is a violation of their sovereignty, but in truth they have facilitated these strikes in return for having accepted payments for them. The combined negative potential of leaders of two nations is utilized to kill people...Available to be killed as sitting ducks...Without even a semblance of a trial.

It puts the common man in America — particularly the sensitive type — in a terrible predicament. He has to carry the stigma that he belongs to a nation whose leaders are engaged in war crimes and it makes it even worse for him when he realizes that he is paying for all this with his taxes.

It puts the terrorist further and further away from becoming a normal human being. Once he "was" indeed a normal human being...with a mom and dad, brothers and sisters...and then became a victim of time caused by the interactions of the world...Poverty, no job, no food. He becomes a paid employee of a company, got trained by a senior also a paid employee. It does not require a solution of complex differential equations to understand that even Osama bin Laden was an employee of the company...perhaps the Chief Operations Officer. He was definitely not the owner. Who are the real owners? Who are the financers? It's astonishing...how they remain elusive... these company owners...Perhaps their total number could be just in double digits or maybe even single digit...perhaps I am wrong...what do I know?

All I know is that these misunderstood human beings are paid employees who could easily have been persuaded to come back to regain normal human-being status with some well-designed words of wisdom and some well-designed guarantees for their safety and security...Why will they not agree to this? What do they lose? But I guess this is an imperfect world...Nobody wants to help them...And with each Drone strike these prospects—of appeasing them—are diminishing. They are now clear who their enemies are and what they are supposed to do next. It's all cause and effect. The truth is we don't give them too many options. Their only hope of salvation...becomes suicide bombers...Kill some more...get some job satisfaction...And get out of this rotten world...What else is there to do?

Now the big question: For whom does this all work out well?

For the Presidents of the United States (POTUS) of course...

(He has no choice but to carry on the work started off by his predecessor (the Republican guy)...He thinks he is successful... effective policeman...never mind "selfish and arrogant"...effective in staving off terrorism from America...How long must it go on? It's already over twelve years since 9/11...How many more drones?

Each strike must be a setback for the company owners...Do they get replacements easily? A question of demand and supply... A news item says they even advertise for suicide bombers...What in the world is going on? It must be a country...it must have leaders selected by its people who want peace...it must have law and order...with all this, is it possible for such ads to appear openly? Are there people so "not taken care of" by their "leaders," who are ready and willing to apply for such jobs?

It's the same old natural selection virus called "to live and let die" that enters minds and creates wars and interferences that has infected some minds and created companies, and now these companies are selecting normal, innocent human beings and injecting this virus into their minds and bodies in such a way as to create a different virus called "to die and let die." (6-026)

Q: What is at the back of the problem in Pakistan and in the Middle East?

SK: Religion...I guess...but actually it is not the real problem... Religion is just an excuse for exploitation...by the company owners.

From what I understand there appear to be many factions of Muslims...They are quite uninformed about each other despite living together...which is very surprising. They have the same God...They follow the same religious document...Perhaps they happen to use different hemispheres of their brains to interpret the content of the document. The most glaring ones pertain to the relation between man and woman. In some factions, there is equality and respect, understanding and love in good measure between the two genders...And in some other factions, not much respect, not much understanding, not much love, and in some extreme cases disrespect.

Other examples pertain to tolerance/intolerance of other religions. In some factions there is tolerance in good measure, while in others, little tolerance of other religions, diminishing in the extreme case to zero tolerance…which means people of all other religions must perish.

But as I said earlier…I doubt very much if religion alone is at the back of all the problems…It doesn't make sense to me that a man x can go out and kill another man y just because x considers his religion to be superior to the religion of y…I repeat it doesn't make sense at all. It's an out-an-out exploitation in the name of religion.., an out-and-out exploitation of the fact that x does not understand these things…let alone understand the essence of religion. "x" is just doing a job for which he is paid and if he dies in the process of doing his job…he is assured…by his bosses of course…of a permanent abode in heaven.

Q: But then…What do the company owners…the financers… gain from all this. Is it merely revenge? And if they — the financers — are only interested in revenge, how is it that they remain elusive…and hidden from law? Or is there something beyond that…perhaps something related to a certain network?

MOC: Refer to the document. There must be beautiful minds in the world to analyze the problems. Leaders of nations cannot solve them…especially the American leadership, unless and until it takes care of that "network."

"Q: What exactly is that "network" ("!") (6-027)

MOC: Refer to the document

Q: So, in summary, there are three problems…Disrespect to women plus Religious extremism plus American interferences… Is that all?

MOC: I guess you could add three more…Population explosion plus Nuclear Arsenal plus Neglecting our environment.

Q: How do we deal with them?

MOC: Peace marches, spearheaded by women plus supported by men plus controlled and safeguarded by security plus well publicized and encouraged by the media should take care of the first two.

Q: What about the third?

MOC: These — American interferences — are of two kinds

a) The "selfish and arrogant" kind meant for its own benefit.

b) The "policeman" kind meant to help other nations in resolving conflicts.

The former must be stopped at once...and the latter may be dealt with by the United Nations through a world force.

Q: Is there a solution to the Middle East problem?

MOC: Finding a solution to the conflicts in the region is as difficult as understanding quantum physics. Indeed the solution looks extremely difficult for the simple reason that...No one...I mean no one...is actually trying to find one. They are all busy predicting... predicting...predicting...and nothing but predicting what's going to happen...by making some assumptions and then applying the law of causation...to the extent they can understand it.

A comprehensive chaos exists in these countries, caused by an extremely complex flow of viruses. These countries need help, like any other country, the common man in these countries too wants to live in peace...And so appropriate antiviruses must be designed to take care of the problems.

Q: What exactly are these Viruses?

MOC: 1) The "Ideology" virus...based on an imperfect understanding of their religious document. Intolerance of other religions…Extremely sharp gender inequalities and consequent disrespect to women are some of the ingredients of this ideology. We all know who these people are…there is no need to elaborate further…In essence they are not religious at all…religion is just a pretext to allow them to take the law in their own hands.

2) The "dictatorship" virus…Where the dictators have links with the "Network"…via a substance called "oil," and now this linkage is understood by the common man and there are revolutions leading to civil wars.

3) Other viruses such as "the hatred toward the West plus Israel" virus, the "free availability of arms" virus, the "nuclear arsenal" virus, etc.

No doubt, it's an enormously complex problem and requires great analytical skills and a substantially collective approach to arrive at a meaningful solution, which is acceptable to all stakeholders in the business. But, there is also an inherent simplicity about it, which is an understanding that the desired end result should be "peaceful coexistence" with each other in the region as well as with the rest of the world.

The best approach would be to work backward from "the desired end result status" and evaluate in comprehensive detail the complete sequence of events and interactions that have the greatest probability to lead toward that "desired end result" that is acceptable to one and all.

Now I do not believe the people in the region are not intelligent enough to understand the problems relating to "ideology"… they can drive their cars, they can use their cell phones…even the latest complex ones, they can use their computers/laptops and operate their music systems, and so on and so forth…This much intelligence is enough to understand that the only "religious" requirement to achieve the end result of "peaceful coexistence" is the will and desire "to live and let live," and that this has to be the common denominator in all the various ideologies.

"A uniform understanding and a uniform and easily understood interpretation of their religious document…An understanding that is commensurate with "peaceful coexistence," "gender equality," and "tolerance of other religions."

With all this…the "trump card" solution to the problems lies in the getting together of all the factions…all the ideologies…and form a collective…"embrace each other"… "practice trade and business with each other"… "work and play with each other"…I think there is a good chance that it would be better than to…"Fight each other."

In short…Let there be a single large country…call it "The United States of the Middle East"…With a single currency (like the Euro zone)…And a single army only to control law and order…with no need for enemy countries.

Other large nations of the world including the United States (notwithstanding the fact that it was once the troublemaker) are bound to come forward and help…Buy not as "policemen."

Q: What is the ultimate solution?

MOC: Design..Determine…and Implement: "the sequence."

Q: What exactly is going wrong with human civilization?

MOC: The second law of thermodynamics is being played out to perfection by the human race. The entropy (disorder and chaos) of human civilization is increasing all the time.

A serious understanding of the law of causation is required to understand what's going wrong with the world and what should be done to prevent the human civilization from moving toward self-destruction.

In particular—as I said earlier—The entropy of the Middle East is extremely high. The situation over there has become so very complicated; it is possible each and every player (leaders of groups) is many times scoring own goals. The biggest contributors to this rise in entropy are: 1) the enormously wide variations in the ideologies and beliefs of the people in nearly every aspect of life, not just religion, 2) The extreme disconnect between the leaders (in most cases dictators with links to the American leaderships) and the common people, and 3) American interferences in the region predominantly for their own benefits. The extent of complexity in the Middle East can be gauged from the fact that sometimes it is possible that the United States and al Qaida are on the same side of the net, as may have happened if the United States had attacked Syria in September 2013 in response to Assad's use of chemical weapons against his own people — the rebels.

Q: Can you elaborate on the American interferences and who are the beneficiaries?

MOC: It looks as if the American leadership sooner or later gets itself entangled in a certain "network." At first I thought that the DOD (Dept of Defense) of the MIC (Military Industrial Complex) of the USA, was an RWS (Republican Welfare System) (6-028). But now I am thinking that even the Democratic POTUSes (6-029) get drawn into it. They start with an intent to control the network, but pretty soon, nothing is in their hands any more. The "network" is in full control over them. Either this or "they get simply out of their minds."

Q: But who controls the "network?" Is it the CEOs? And who finances the "network?"...Is it the American taxpayer's money? But that should bring down the economy...Or is Switzerland involved?

MOC: This is too complex for me.

Q: What does the common man in America feel about this?

MOC: The common people (especially women) in America are dead against this. Americans by and large are sensitive and intelligent (in some respects), and are absolutely dead against American interferences in the affairs of the rest of the world. But they are helpless. In America too, as in many countries of the world there is a sharp disconnect between the political leaders and the common people.

Q: What else is leading to the rise in entropy?

MOC: The virus of "Corruption in High Places." This is rampant in nearly every country, some of these countries — including India — are specializing in this.

Q: What is the antivirus for this?

MOC: Perhaps..."The virus must become its own antivirus and annihilate corruption."...Just as the quarks annihilated the antiquarks during the Big Bang.

Q: Is not that too far-fetched to expect?

MOC: Maybe it is...Maybe not...It is a huge opportunity — provided by the "six words" to become heroes overnight.

The Penultimate Question:

Q: How should the world be educated on all these things?

MOC: "Survival from collapse" must become the most important subject to be taught in the universities worldwide.

And the last question:

Q: What is our trump card?

MOC: The concept of an "infinite mind "and "The Six Words."

**

(6-030)
"Yes there has "always been something or someone in my head" and it wasn't me.
And now I can sing:
"I saw the light...I saw the light...no more darkness...no more night...now I am so happy, no sorrow in sight...Praise the lord...I saw the light." Hank Williams

ENDNOTES (6-023 TO 6-030)

6-023:

Extract from Roger Penrose's book...*The large, the Small and the Human Mind*

6-024:

Extracts from Charles Townes's views on evolution and intelligent design during an interview. (http://berkeley.edu/news/media/releases/2005/06/17_townes.shtml)

6-025:

Extract from James Martin's book...*The Meaning of the 21st Century*

6-026:

For further reading, refer to yours truly's response to a news blog "Pakistan...Drone Strikes are Violations of Sovereignty" on huffingtonpost posted on June 4, 2012.

6-027:

MIC = The military industrial complex (of America)

6-028:

For further reading refer to yours truly's response to Howard Fineman Blog "Romney, in RNC Speech Short on Specifics, Fails to Close the Deal" posted on August 31, 2012

6-029:

POTUS = President of the United States (of America)

6-030:

Six words — Twenty letters — Twenty Quotations/Statements

**

EPILOGUES

EPILOGUE ONE

In yet another parallel universe the question-and-answer session still continued:

Q: How do we deal with the misunderstood human beings? Will it be appeasement…or will it be…?

MOC: The MHBs are not at fault…they are just paid employees…victims of time…they are welcome to come back and live with the NHBs…it's "the financers," the "company owners" we need to deal with.

Q: How do we deal with the…"company owners?"

MOC: Refer to the document.

Q (From a lady delegate): Will it be a good idea to invite a few of them…to the next seminar?

MOC took her time to answer the question…there was silence… music…silence again and the delegate asked the question again for the second time…

Q: Will it be a good idea to invite a few of them…to the next seminar?

MOC: We invited them to this one…not just "the financers" …or "the company owners"…we also invited…"the tainted leaders of the world including the leaders of the MICs of the world." not forgetting…"The account holders"…etc.

The lights went out…The curtains came down…The music kept playing…

The tone changed…The lights…of a different color…were switched on one by one…The curtain rose again slowly.

Six of them—the invited ones—appeared on the stage… flanked by the other two ladies…Believe it or not…there was an applause to welcome them.

The lady delegate: Great job…MOC!

And then…one of the six came forward and took center stage…and gave his presentation…He was a changed man…and so were all six of them who represented "the tainted ones."

A brief extract from his speech:

"Even in our wildest of dreams we could not have imagined that we would get an opportunity like this to explain to the world why some of "us" did things we should not have done.

Past history of at least the last seventy years or so, along with a serious understanding of the "interactions theory," must be taken into consideration in order to understand why some of us did things we should not have done…and then deciding on the course of action that the world leaders must take to ensure future security of the human beings.

It is true, we, the human beings, are the outcome of natural selection and to the operation of this law can be attributed the fitness of the human body and brain…but this law cannot be operated by improving the fitness of one section of human society at the cost of causing suffering to another section. The evolutionary game must change…Interference is not the answer…global approach and cooperation is the answer.

It is true "ours" (a part of "us") was a wrong hand and we were on our way to do things…that we may not do now…as a consequence of having understood the "interactions theory" and the "infinite mind"…but I would like to say that the:

"responsibility toward the future security of the world lies not just in preventing wrong hands from getting powerful but also in preventing powerful hands from possible wrong actions."

One by one all the six tainted leaders of different varieties... took the stage and first explained themselves and then came forward and gave suggestions.

The consensus view that emerged was...that "The Three Divides" must be brought to a minimum...the divide "between the rich and the poor"...the divide "between the West and Islam"...and above all the divide "between the various factions of the Muslim World."

The consensus view that emerged was...that the problems of the world that needed to be dealt with are...deprived societies... rogue governments...dispersed militias...corruption in high places...rampant thuggery, etc.

The consensus view that emerged was...that the nuclear arsenal of the world must be destroyed as early as possible.

The consensus view that emerged was that the troublemakers themselves must now become the troubleshooters...*for now they know what the six words are.*

And then there was a group photograph...The "MOC" joined the group of six in the front row...and the twenty-five speakers in the second and third rows...and all the delegates were in the rear...The two other ladies of the organizing group took the picture.

And the caption was:

** *** **** *** **** ****

EPILOGUE TWO

A couple of days after the seminar...A news blog covering the seminar was posted on the *Huffington Post*.

Plenty of thought provoking responses/comments ensued...

Noticed some intriguing ones like:

Bellanova: Who were these tainted leaders?…and why did the MOC chose to sit in the front row with these people in the photograph?…who did she represent?

One easy rider: Depends on which parallel universe we are in.

**

EPILOGUE THREE

In another one…

Q: What is the ratio of matter to emptiness in the universe as a whole?

A: Five atoms per cubic meter of space will stop the expansion of the universe. But the universe is expanding, so we don't have five atoms per cubic meter. Available data shows that there are only about 0.2 atoms per cubic meter of space. That would make the expansion very fast. But the universe is not expanding that fast. The current rate of expansion of the universe shows the amount of matter as equivalent to about four atoms per cubic meter. But the available data shows that the accountable/visible matter as only 0.2 atoms per cubic meter, which means the remaining matter equivalent to about 3.8 atoms per cubic meter must be some unknown type of matter such as either dark matter or dark energy.

For all practical purposes we can say that the total amount of matter does not exceed four atoms per cubic meter of space.

Q: What then is the ratio?

A: One unit of matter in a trillion trillion units of void…And that's not all. In each each atom the nucleus comprising of protons, neutrons, with orbiting electrons, make up just an insignificant part of the atom, like a few grains of sand in an auditorium.

Q: What then is the ratio?

A: One unit of matter in several trillion, trillion, trillion units of void.

**

Q: In the video clip shown on the second morning…in the conversation between Eugene Wigner and John Wheeler…Wigner asked a very interesting question of Wheeler…the last few lines of the question were: What is a similar problem in the black hole situation? What does become uninteresting? Tell us, John…What did Wigner mean when he said, "What does become uninteresting?"

MOC: I'll let SK answer this

But SK had no answer…

A few more questions…

But SK had no answer…

Charles Townes: What's the matter…SK?

Indeed, SK had no answer…

Then a sixth sense told him…He should look for something…

He looked out through the window…and saw the UFO moving away…and receding from vision…(6-013 of SIX B)

And he knew…There was indeed someone in his head and it wasn't him.
And now that someone has left…and he was back to square one…and he had to confess…
"I know now that I know nothing."

**

Roger Penrose: He means, "whats nothing."

Paul Davies: Or maybe "the infinite mind."

A UNIVERSE OF PURE THOUGHT

ACKNOWLEDGEMENTS

To begin with I acknowledge with deep thanks the great scientists who inspired me. I am specially thankful to those whose philosophical viewpoints enabled me to arrive at my `Standard model of philosophy'. To name just a few, they are Albert Einstein, Erwin Schrodinger, Charles Sherrington, John Wheeler, Eugene Wigner, Roger Penrose, Paul Davies, Charles Townes, etc.

I am deeply indebted to Huffington Post for giving me ample opportunities to give my comments and engage in healthy discussions with bloggers and other HP users on topics related to the subject of this book. These discussions were enormously helpful to me in formulating the manuscript of chapter six `The Seminar`. To name just a few of these friends on HP, they are Dave Astor, Louis Del Monte, Peter Baksa, Diogenes of Alaska (not real name), Oneeasyrider (not real name), Jill Press, Bellanova (perhaps not real name), Charlie foxtrot999(not real name), etc

I am also indebted to several of my friends and colleagues, for having read the manuscript, for supporting my philosophical viewpoints, as well as for their appreciative views (not forgetting critical ones) and suggestions. This is a very big list. To name just a few, they are, Mr. Vivek Sinha, Dave Astor, Dr Clifton Meador, Sumit Chowdhary, Dr. A.K. Jain, Shoumen Datta, Anindya Maitra, Joe Fernandez, Bijou Chatterji, Begum Shaherbanoo Lagad, Dr. Raja Reddy, Prakash Newalkar, the first three named are also writers themselves, and I thank them for the `wise counsel` I received from them.

Wise counsel was also received frequently from Dr A.K.Jain, Professor and Head of the Physics Dept at the Prestigious I.I.T

Roorkee, who became a close friend after we met at the 2003 Symposium on Science and Beyond held at Bangalore, India. He was one of the Speakers at the symposium attended by some of the greatest scientists and philosophers of the World. When last we met, recently, he invited me to give a talk on Science and Philosophy at his institute. This is an honor for me, and this opportunity will be difficult to resist.

Special thanks are due to Sumit Chowdhry for reading the complete book word for word, not once, but twice, for his reviews and for offering valuable inputs and constructive suggestions. We had several rounds of lengthy interactions before incorporating these in the manuscript. Special thanks are due to Anindya Maitra for his comprehensive review of the book. The inputs received from Sumit Chowdhry and Anindya Maitra were immensely helpful to me in editing. I also thank Shoumen Datta for his review, in particular for his critical review, which was quite helpful in trimming the manuscript and making it look cohesive.

Thanks to my lovely artist wife `Bharati` for her unstinted support and for agreeing to make four paintings on the subject for my book launch. Thanks also to my Sons Kamal and Nikhil, daughters in law Shibanee and Mary Kathleen, niece Vinita Sahi, Dr A.P.Handa and Sudhakar Hannda for their help and support.

Special thanks are also due to Mr. Vivek Sinha for his valuable guidance in matters relating to Publishing, and to Dr. Clifton Meador for informing me about `Create Space` and suggesting I should go with them.

And of course, thanks are due in great measure to `Create space` for copy editing, as well as for the interior and cover designs. I am extremely pleased with the cover design which is exactly as I wanted. The copy editing carried out by Create Space team was brilliant and comprehensive, specially the comments and suggestions given, and the questions asked, that provided me with ample opportunities to make improvements where required, and to provide clarifications where I thought I was right. Indeed, the decision to go with `Create Space` was a vital one, and gave me much happiness.

Special thanks to son Nikhil for his interactions and perfect coordination with `Create Space`, and to both Nikhil and MK for their valuable suggestions in several aspects relating to the book.

Last but definitely not the least, I have to thank all my colleagues at `Total Environment Building Systems Pvt Ltd`, with whom I Interacted vigorously in connection with the book. These philosophical discussions, going on for several years now, have given me immense pleasure. I have a feeling these interactions were helpful in the development of team spirit and camaraderie in the organization. This is a very big list ...above fifty...not enough room here. I shall thank them personally, and soon.

AUTHOR BIOGRAPHY

Surendra Kumar Sagar earned a BS from Bombay University with a focus on math, physics, and statistics, as well as a BS in civil engineering from Thapar Institute of Engineering and Technology five years later.

With over fifty years of experience in the design and construction business, Sagar presently works as the technical director of design and engineering for Total Environment Building Systems and is a member of the Institution of Engineers India.

Sagar currently lives with his wife in Bangalore, India. They have two sons and three grandchildren.

PRAISE FOR THE BOOK:

``Six Words is an autobiographical epic story – epic in the dictionary sense to mean "extending beyond the usual or ordinary, especially in size or scope."

Sagar begins his personal journey at the Big Bang origin of the universe with "I am a quark". He progresses to become an atom of hydrogen, then helium, and finally explodes out of a supernova toward earth as a carbon atom as our planetary system has formed. On earth be becomes an organic molecule and after millions of years he finally becomes a structural engineer.

This book is a tour de force of the major physical sciences, theology, and philosophy. Sagar goes deeply into each, explaining in clear terms very complex subjects. The book then moves to a hypothetical seminar in which the major scientists and philosophers gather to compare notes and thinking. Einstein is there along with Erwin Schrodinger, Charles Sherrington, John Wheeler, Eugene Wigner, and many other top scientists from history.

This is a major five star book written by a serious student, thinker, and observer of the sciences of the universe. The book's aim is to draw religion and science together in a way that leaves established scientific laws and rules intact.

The Six Words do exactly that. I will leave it to the reader to have the joy of uncovering Sagar's Six Words.``

By... Dr. Clifton K Meador...Author of ` Fascinoma`

247

``In just three lines one can summarize the book "six words" by Sagar. It opens your mind to an unknown territory; it introduces you to something that has been the binding factor of the universe but we seldom thought about it, and finally it empowers to explore the complex philosophical questions which are generally discussed without any formal conclusion.

The book helps you understand and dissect scientific knowledge in three levels — First it introduces you to basic science and makes you comfortable by telling "you know what you know" then at the second stage it slowly and politely tells you that "you also know what you don't know" and finally, a large portion of the book tells you "you don't know even what you don't know".

it opens door to a new world, it navigates you to your seventh sense, which you never used and tells you what miracles a human kind can do.

It is ultimately an exploration into how we humans can come together and change the future course of mankind's history.``

By N Raghuraman, Author, ex editor
(Dainik Bhaskar, DNA and Indian express)

``I am reminded of Stephen J Gould`s words…"If genius has any common denominator, I would propose breadth of interest and the ability to construct fruitful analogies between fields." That is what Sagar has accomplished in his book ` Six Words`.``

By Showmen Datta. Director MIT,
Academician, Author, Research Scientist at MIT

``Written with great passion and depth, is very readable by the 'science-minded' as well as the non-scientific. Exploring the corridors of religions, beliefs, philosophies, history of the world to a backdrop of quantum physics, the author takes you through a thought-process so tremendous in its global dimensions that you must pause, deliberate and 'know' that now is the time to seriously consider that next step to human and global preservation.

A must read for all who have any consideration for the future of this earth and its inhabitants - some your own progeny!``

By Sudhakar Hannda ....on Amazon

``Years of inspiration soaked from Quantum Physicists, leads an engineer into a state of mind that makes the writing of this book the inevitable outcome. Served on a platter is a capsule potent with the author's own years of assimilation of the counter intuitive ideas that form the backbone of Quantum Physics. Look no further as Science and Philosophy lose the fine line that divides these two disciplines and engages with a deeper you and makes you appreciate the very 'practical' solution to the most fundamental problem - the continued existence of the Human Race.

A roller coaster ride through the most unique autobiography that one is likely to come across - an autobiography that starts at the Big Bang...``

By Sumit Chowdhury ...on Amazon

"...a treatise that ponders the laws of physics, the history of the cosmos, the nature of God and the fate of mankind.
"...kind of "autobiography" of his existence, starting with the formation of his constituent subatomic particles..." "... It begins with a brief, engaging account of cosmology from the Big Bang through the evolution of life. The book then turns to more involved (and less successful) explorations of advanced physics, including the mysteries of Heisenberg's uncertainty principle, "quantum entanglement" and the relativistic paradoxes of travel near the speed of light..."
"... The book's sixth chapter comprises a fanciful "seminar" of great thinkers—from Immanuel Kant to Albert Einstein to contemporary physicist Freeman Dyson..."
"... All this background sets up a section on Sagar's own philosophical speculations, which mix such topics as the anthropic

principle—which says that fundamental constants must be able to support the life-forms that observe them—with the quantum mechanics mysticism..."

"...Sagar theorizes that God is an abstract "all intelligent omnipresent...infinite mind"; that humans may eventually merge into the divine "Super consciousness"; and that our main task is to avoid blowing ourselves up in the next few centuries—a disaster that Sagar considers a near-certainty unless everyone works for world peace.``   ...   ...   KIRKUS REVIEWS

``It is not that I wrote this book ``Six Words``...It is that it was written by me...consequent on the interactions of the world on me. I even do not know...if this ... the writing of the book is attributed mainly to the revolving doors of chance, or to the existence of a certain ``intelligent field``, that   has created   probabilities that books of this kind should be written to prevent  the human civilization from getting extinct. There is also an ``Entropy field`` which is continuously increasing disorder and chaos and leading us towards self destruction. We do not need to solve complex differential equations to understand and realize how far the `intelligent field` is trailing behind the `entropy field`. Will the former be able to overtake the latter before it reaches its winning post of destruction?. Its entirely left to us human beings...No God of any kind will come to our rescue.

...Author`s own confession
`` Damn          Good          Advice``
By .... George Lois .... Author of `Damn  Good  Advice`